TEACHING STRATEGY:
CHALLENGE AND RESPONSE

Gabriel Marcella
Editor

March 2010

*****

CONTENTS

FOREWORD

No subject is more essential in the preparation of national security professionals and military leaders than the teaching of strategy, from grand to military strategy. Nor is there one that is more timeless and intellectually demanding. Moreover, the experience of the armed forces in recent wars recommends that the system of military education needs to conduct a serious analysis of the way strategy is taught. The task is even more imperative because the ambiguous conflicts and the complex geopolitical environment of the future are likely to challenge the community of strategists, civilian as well as military, in ways not seen in the past. In this context, developing the appropriate curriculum and effective methods of teaching strategy will be the responsibility of universities, colleges, and institutions of professional military education.

The authors of this compendium ask and answer the central question of how to teach strategy. The findings, insights, and recommendations are those of professionals who are accomplished in the classroom as well as the crucible of strategy. This book should stimulate discussion and introspection that will in time enhance the security of our nation.

The Strategic Studies Institute is very pleased to publish this volume as an important contribution to the education of our future leaders.

DOUGLAS C. LOVELACE, JR.
Director
Strategic Studies Institute

CHAPTER 1

INTRODUCTION

Robert H. Dorff

THE DEBATE ABOUT TEACHING STRATEGY

The general topic of this book became part of a very public debate in the United States in 2009, and it continues today. That debate concerns how one should teach strategy in our system of professional military education (PME). The genesis of the debate was alleged shortcomings in strategy in the wars in Iraq and Afghanistan, or what some critics have called failures in strategy and strategic leadership. The allegations subsequently led some members of Congress to turn their attention specifically to the PME system since that is where our military leaders supposedly learn their skills in formulating and implementing strategy. And so the question: Is there something wrong with the PME system? A virtual parade of experts has testified on various aspects of this question before special hearings conducted by the House Armed Services Subcommittee on Oversight and Investigations.[1] In some circles, a presumption existed that the PME system had failed this country by failing to educate senior leaders on how to produce sound and effective strategy. According to this line of reasoning, those senior military leaders were at least partly responsible for the strategic shortcomings or failures. While our military strategy could seemingly win the war as it was fought on the battlefield, we could not achieve overall strategic objectives. Since those objectives comprise the very reasons for which a war is fought, was this not indicative of a fundamentally flawed strategy? And if so, should we not examine the system of education that is supposed to produce strategic leaders, especially in the ranks of our senior military officers? Almost 1 year into this formal examination, it remains to be seen whether a review of PME and the debates it has spawned will yield any real changes in either the teaching of strategy or, most importantly, its practice.

THE ENDURING DEBATE ABOUT TEACHING STRATEGY

It would be a mistake to think that the public debate is either a reason why this book appears now or that it is the only debate on the topic. While the contributors to this volume do shed light on a number of the important questions raised as part of that public debate, the simple fact is the authors (and many others) are part of an ongoing debate that has deep roots. The essays in this book were first presented formally at a workshop conducted at the U.S. Army War College (USAWC) on April 17, 2009, following the 2009 USAWC Annual Strategy Conference. At that "Teaching Strategy Workshop," authors presented papers in panels and then opened the floor to discussion with an audience consisting of both PME and civilian academic faculty, as well as military and civilian practitioners of national security affairs. Of course, by that time, almost all of the participants at the workshop were aware of the congressional hearings begun the month before. But even the origins of the 2009 workshop and the various lines of thought presented there went further back. A similar workshop on the same topic was organized by the Strategic Studies Institute (SSI) the year before. In fact, the presentations and discussion that were part of that April 2008 workshop were even then merely a reflection of a debate that had started up again in 2007. I say started up again because anyone familiar with PME generally, and the Senior Service Colleges (SSC) specifically, knows that questions of what we teach and how we teach it are recurring themes. Those questions as they relate to strategy may die down from time to time, but they never really go away completely. They are frequently rekindled either by events on the ground, or by new analysis and thought, or some combination of the two. The point here is that what the reader will find in the chapters that follow is not simply a reflection of the current debate about PME, but part of a long-standing and ongoing search for ways to improve our understanding of what strategy is and how best to incorporate it in the development of our future leaders.

THE BOOK

Not surprisingly then, some of what is contained in this volume addresses PME and the senior service colleges directly. Some of the authors in fact focus specifically on the curriculum at the institution where they teach or have taught in the recent past.[2] At least one author addresses some of the same questions from the perspective of a civilian academic "who teaches about strategy at a liberal arts college."[3] Others choose to emphasize teaching approaches, techniques, and concepts or broad issues dealing with honing conceptual and thinking skills that may be required in the increasingly complex and ambiguous realm of strategy formulation and implementation.[4] One essay focuses explicitly on the relationship between how we define strategy (the "what is strategy?" question) and the way we subsequently design a curriculum to teach concepts and skills based on that dominant paradigm of how strategy is defined.[5] But one singularly important question lies at the core of this entire volume: Do/Can we teach not just an *understanding* of strategy but the ability to *do* it? In other words, can we actually better prepare students to formulate and implement strategy, and if so, how?[6]

Since this is an edited volume on teaching strategy, there will be many common threads running through all of the chapters. Readers will find a number of recurring questions: What is strategy? Why should we teach it? What should we teach? How should we teach it? What should we expect as reasonable and necessary outcomes? While there is considerable agreement in principle on answers to some of these questions, not all of the authors see things the same way. Some of that is a function of different perspectives brought to bear on the questions; some is a function of more basic disagreement about the essence of strategy and effective pedagogy. But that is as it should be if we are to continue two very critical processes in our roles and responsibilities as teachers: Learning how to learn and learning how to teach. For many years now, I have taught that strategy is best understood as a dynamic process in which a continuous set of multiple interactions taken together generate outcomes: Interaction among individuals and the choices they make; interactions between individuals and the often-shifting strategic environment; and interactions occurring on a highly fluid playing field with different players entering and exiting the "game" at different times, just to list a few examples. But those outcomes are themselves never fixed as the iterations of play continue—hence the dynamic process that has no discrete termination point. Therefore, what is required first and foremost for strategic success, beyond the necessary understanding of the nature of strategy, is the recognition that strategic thought, action, and leadership will always be in play and subject to ongoing assessment, reassessment, and adjustment. Is it therefore not only fitting but fundamentally logical that teaching strategy—the "what" we teach and the "how" we teach—should also be part of a similarly dynamic and constantly evolving process? I think so, and I suspect that the contributors to this volume share that same view. It is in the spirit of contributing to this ongoing process of assessing, reassessing, adapting, and adjusting that the authors offer their thoughts on how best to teach strategy.

Today, of course, there is a near surfeit of published work arguing that in the early years of the Iraq war the United States failed to formulate and implement a sound strategy. That part of the premise on which the congressional hearings were based is almost undeniable. For example, writing in 2007, Steven Metz concluded that "[H]owever laudable the overarching American objectives in Iraq, the United States was strategically and conceptually unprepared to realize them. We used flawed strategic assumptions, did not plan adequately, and had a doctrinal void. . . . American strategy was characterized by a pervasive means/end mismatch."[7] Why was that? There are many possible answers, and almost all of them have proponents somewhere. Most of them fall into two broad categories: Shortcomings of our national security processes and organization, and shortcomings in individuals. Of course, some analysts point to combinations of the two. One recent study frequently cited today comes down strongly in the second category, concluding: "The deeper problem seems to be more a lack of understanding of what strategy is than structural or organizational defects in the United States' national security establishment . . . US political and military leaders have been increasingly inclined to equate strategy with listing desirable goals, as opposed to figuring out how to achieve them."[8] Of course, even if this is in fact not the primary cause of flawed strategy, the likelihood of individuals redesigning or reorganizing effective processes and organizations for formulating and implementing strategy is bound to be miniscule if those same individuals lack an adequate understanding of what strategy is. For that reason alone, it appears that the overall focus of this book is highly relevant and well worth addressing.

Were the most recent U.S. strategic shortcomings a function of failures in the PME system? I will leave that question for others to answer. Is it time to revisit the questions raised throughout this edited volume, and elsewhere, about PME and the curricula not only at the SSC-level, but at all levels? The answer to that question is in my view quite simple, albeit rather banal: It is always time to do that. And as several authors argue here, it is also high time that as a society we raise a similar concern about what is and is not being taught, and how it is taught, in our civilian institutions of higher learning. There is little doubt about the desirability of an informed citizenry. Two reasons stand out in terms of the topic addressed here, and the need for sound strategy. First, the public is a key player in the democratic processes in which strategy is both formulated and implemented. And second, the public comprises the pool from which our future political leaders will emerge, the civilian leadership who will still exert primary control over the formulation and implementation of strategy. Therefore, it is not just the military strategists and their knowledge and skills to which we should attend; civilian strategists are just as sorely needed. The questions dealing with teaching strategy—why we should study it, what we should teach, and how we should teach it—may bear most directly on the system of PME. But the answers need to be applied much more broadly across a wider range of our society today. For only then can we expect to regain strategic competence, not just in the crisis of the moment but in a sustained manner well into the 21st century. The contributions to this edited volume will advance that society-wide discussion and debate.

ENDNOTES - CHAPTER 1

1. Testimony before this committee is available from *armedservices.house.gov/hearing_information.shtml*. Testimony on this topic can be found on March 19, May 20, June 4, and September 10, 2009, to list just some of the dates. Those testifying included Andrew Krepinevich, Dr. James Jay Carafano, Major General Robert Williams, Rear Admiral J. P. Wisecup, Lieutenant General (Ret.) David Barno, and Dr. John Allen Williams.

2. Although all of the essays in this volume go well beyond a simple explication of a particular SSC curriculum, this group includes Marcella and Fought, Chapter 4; Lee, Chapter 5; Watson, Chapter 6; and Yarger, Chapter 7.

3. Gray, Chapter 3 of this book, p. 47.

4. Kennedy, Chapter 2; Franke, Chapter 9; Harrison, Chapter 10; and Paparone, Chapter 11, are examples.

5. Da Costa, Chapter 8.

6. In a monograph published by SSI after this edited volume was assembled, Colin Gray emphasizes this theme in *Schools for Strategy: Teaching Strategy for 21st Century Conflict*, Carlisle, PA:

Strategic Studies Institute, U.S. Army War College, November 2009. He observes: ". . . both the theory and the practice of strategy need to be taught, insofar as they can be, because an education in strategy must encompass ideas and the application of those ideas as plans that have to be implemented by command performance." (p. 7) Also available from *www.strategicstudiesinstitute.army.mil/pubs/display.cfm?pubID=947*.

7. Steven Metz, *Learning from Iraq: Counterinsurgency in American Strategy*, Carlisle, PA: Strategic Studies Institute, U.S. Army War College, 2007, p. 85.

8. Andrew F. Krepinevich and Barry D. Watts, *Regaining Strategic Competence*, Washington, DC: Center for Strategic and Budgetary Assessments, 2009, pp. vii-viii.

CHAPTER 2

THE ELEMENTS OF STRATEGIC THINKING:
A PRACTICAL GUIDE

Robert Kennedy

With 58,000 American lives lost, 350,000 casualties, and untold national treasure forfeited, on April 30, 1975, the last Americans in South Vietnam were airlifted out of the country as Saigon fell to communist forces at the height of the Cold War. A few days earlier, with the end in clear view, a Four Party Joint Military Team, established under provisions of the January 1973 Paris peace accords, met in Hanoi, North Vietnam. At that meeting, Colonel Harry Summers, Chief, Negotiations Division of the U.S. Delegation, in a conversation with Colonel Tu, Chief of the North Vietnamese Delegation remarked: "You know you never defeated us on the battlefield." Colonel Tu responded: "That may be so, but it is also irrelevant."[1] So was told the story of failed strategy.

It might be facile to contend that the need for systematic thinking about U.S. foreign and security policies and defense issues peaked during the Cold War. After all, during the Cold War the Soviet Union came to pose a military threat to the United States that was unique in American history — the threat of instant annihilation. It also posed a direct military threat to our allies in Europe and Asia whom we were pledged to defend, as well as the danger of ever increasing Soviet influence around the world through proxy wars and other forms of political violence that seemed to some to represent a more subtle, more likely, and perhaps graver long-term threat to the overall security and well-being of the United States. Thus the objectives were clear. First, counterbalance Soviet strategic power and its military might on the continent of Europe with countervailing theater and strategic forces that could deliver responses to any aggression by the Union of Soviet Socialist Republics (USSR) so devastating that no Soviet leader would dare take such a step. Second, contain the growth of Soviet influence through policies designed to thwart attempts by the USSR to subvert governments friendly to the United States and its allies. Though the objectives were clear, the methods to accomplish these twin tasks were not. Here systematic thinking was at a premium, albeit not always wisely undertaken.

With the demise of the Soviet Union and the end of the Cold War, these dangers disappeared. Reflecting this change in environment, funds for intelligence, diplomacy, and defense were initially cut and a number of promising signs emerged. For example, the latest available data indicate there has been a marked decrease in armed conflicts. Notwithstanding Rwanda, Srebrenica, and elsewhere, the number of genocides and political murders has plummeted. The dollar value of major arms transfers has fallen. The number of refugees dropped. And five out of six regions in the developing world have seen a net decrease in core human rights abuses.[2]

Nevertheless, today's world and most certainly the world of tomorrow demand no less in terms of strategic thinking than in the past. The events of September 11, 2001 (9/11) served as a painful reminder that we have not yet reached the end of history, postulated and described by one pundit as that time where conflict is replaced by "economic calculations, the endless solving of technical problems, environmental concern, and the satisfaction of consumer demands."[3] The world has become more, not less, complex. The single great adversary, fixed geographically, is gone. But in its place are multiple threats and challenges, few of which emanate from a single nation-state and few of which seem to pose the immense and immediate danger that confronted the United States during the Cold War. Today, ethnic strife threatens the stability of nations and ethnic cleansing challenges America's most fundamental ideals. Drug cartels and transnational organized crime and their handmaiden, corruption, undermine governments and threaten our economy and the economies of our allies and friends and nations upon whom we depend for scarce resources and/or markets. Trafficking human beings is an affront to our moral values and violates our sense of what the post-Cold War order should represent. Environmental degradation challenges the health of our citizens and future economic progress. These are but a few of the challenges that we must now address.

There are, as well, some challenges, which if not carefully confronted, are likely to pose unimaginable dangers for the United States, its people, and others around the world. Among the more prominent are those resulting from the explosion of technology and technological know-how. Attacks on cyber networks can endanger national political, military, and economic infrastructures, with global implications for the safety and welfare of peoples. The increasing availability of biological, chemical, radiological, and nuclear technologies, which, if acquired by terrorists, so-called rogue states, or perhaps even malevolent individuals, could threaten the very existence of peoples and societies. These challenges are real and demand today, and in the future, careful attention and systematic thinking if we are to preclude disaster. It will require that America bring its domestic resources to bear. It also will require that the United States build partners abroad, both with governments and with individuals ready and willing to contribute to our efforts. It will require reducing the numbers of those who collaborate with or knowingly ignore those insistent on doing harm, and increasing the numbers of governments willing to aid and individuals willing to risk their lives to provide the United States and other governments with information necessary to thwart those with dangerous designs against individuals and nations. In short, the challenges of today and tomorrow will require well-designed strategies if we are to be successful in preserving our values, our institutions, and our nation.

This will not be an easy task. In general, Americans are a pragmatic people. Frequently impatient when confronted with complex solutions to problems they must address, they tend to prefer direct approaches. They are action oriented rather than reflective, a-strategic if not anti-strategic, and all too frequently anti-intellectual, favoring simple solutions rather than the more involved. They prefer checkers to chess and the approaches of Gary Cooper at High Noon and John Wayne to the difficult tasks of examining alternative solutions to complex problems.

French conservative Lucien Romier, writing early in the last century, noted that Americans have a preference for action, for speed or practical efficiency rather than depth, and constant and lightning changes rather than enduring qualities. Writing a few years earlier, Russian political economist and sociologist M. Y. Ostrogorski observed: "Of all the races in an advanced stage of civilization, the American is the least accessible to long views. . . . He is preeminently the man of short views, views often 'big' in point of conception, but necessarily short."[4] Alexis de Tocqueville, in his *Democracy in America,* concluded: "democracy is unable to regulate the details of an important undertaking, to persevere in a design and to work out its execution in the presence of serious obstacles. It cannot combine measures with secrecy, and it will not await their consequences with patience . . . democracies . . . obey the impulse of passion rather than the suggestion of prudence."[5]

Closer to home, Clyde and Florence Kluckholm in their mid-20th century study of American culture contended that Americans believe in simple answers and distrust and reject complex ones. According to the Kluckholms, Americans also tend to be anti-expert and anti-intellectual.[6]

To add to the problem, generally speaking, American colleges and universities do not produce strategists. Outside of business schools, few offer courses on how to think strategically. Even in our senior military educational institutions, the study of strategy often devolves to the study of a few great strategic thinkers, coupled with the study of the national security processes (both necessary, but insufficient), rather than an analysis of what it takes to be a sound strategist. Yet the ability to think strategically is precisely the quality that will be required of America's leaders if the United States is to deal successfully with future problems.

STRATEGY — AN ACTIVITY OF THE MIND

The word *strategy* comes from Greek words *stratēgia* (generalship) and stratēgos (general or leader).[7] Historically, the term strategy has been associated with military activity. The father of modern strategic studies, German Major-General Carl von Clausewitz, defined strategy as "the use of the engagement for the purpose of the war."[8] Field Marshall Helmut Carl Bernhard Graf von Moltke contended that strategy was "the practical adaptation of the means placed at a general's disposal to the attainment of the object in view."[9] Placing less emphasis on the battles, Sir Basil Henry Liddell Hart criticized Clausewitz, contending that Clausewitz' emphasis on battles suggests that battles were the only means of achieving strategic ends.[10] Thus, Liddell Hart defined strategy as "the art of distributing and applying military means to fulfill the ends of policy."[11] Liddell Hart's definition suggests a somewhat wider variety of military means, and clearly emphasizes that the political objectives are the ends to be pursued by military means. Of course, Clausewitz made the latter point early in his seminal *On War* by his famous dictum "war is not a mere act of policy, but a true political instrument, a continuation of political activity by other means."[12]

Increasingly in the 20th century, students of strategy extended the definition well beyond the field of military activity, applying the term regularly in such fields as business, politics, and foreign and security policy. While the Merriam-Webster dictionary, paying partial deference to earlier uses of the word in a military context, provides as its first definition "the science and art of employing the political, economic, psychological, and military forces of a nation or group of nations to afford the maximum support for adopted policies in peace and war."[13] It simplifies but broadens the understanding of strategy, providing it with its modern look, in its second definition: "a careful plan or method; the art of devising or employing plans or stratagems toward a goal."[14] Both definitions miss the mark. In the simplest of terms, strategy is the *integrated* application of available means to accomplish desired ends. The emphasis is on integrated. The first definition misses this important point. The second definition, though perhaps too broad to be useful, does emphasize that strategy is simply a game plan. The haphazard or spontaneous employment of means cannot be considered strategy.

At the national political or military level, a more useful definition of strategy is the integrated application of the instruments of national power (e.g., political/ diplomatic, psychological, economic, informational, and military) in pursuit of national interests. Strategy understood as the integrated application of available means to accomplish desired ends, of course, does not limit strategy to the use only of available means. A well-developed strategy may include efforts that lead to an enhancement of means.

Despite this seeming simplicity, strategy is a term that is frequently misused. For example, during the Cold War the security and defense community often referred to the strategy of containment. Yet strategy cannot be a simple restatement of an objective, such as containment or the containing of the Soviet threat. To do so ignores the fact that there can be multiple avenues of approach to accomplishing an objective. Nor can strategy easily be reduced to a single term. It is a multiplicity of actions, carefully integrating available means in order to achieve desired ends.

Strategy is neither strictly art nor science. Yet, in some ways, it is both. As an art, the ability to think strategically is a skill that can be acquired through experience, observation, and study. As a science, thinking strategically entails the systematic pursuit of knowledge involving, among other things, the recognition and formulation of a problem, the collecting of information, and the formulation and testing/analysis of alternative hypotheses. However, strategy is preeminently an activity of the mind. As was war for Clausewitz,[15] strategy is an act of human intercourse. It is about influencing behavior. It is the formulation of a game plan designed to get inside the decisionmaking loop of others, to get them to do what they might not otherwise have done—whether in the halls of government, in the boardroom, or on the battlefield. So it was for Sun Tzu, who wrote: ". . . to win 100 victories in 100 battles is not the acme of skill. To subdue the enemy without fighting is the acme of skill;"[16] and, "those skilled in war subdue the enemy's army without battle. They capture his cities without assaulting them and overthrow his state without protracted operations."[17]

Reflecting a similar thought, Tu Mu, writing sometime between 619 and 905 A.D., observed: "He who excels at resolving difficulties does so before they arise. He who excels in conquering his enemies triumphs before threats materialize."[18] Nearly a millennium and a half later, in a note to himself, Liddell Hart wrote: "to influence man's thought is far more important and more lasting in effect than to control their bodies or regulate their actions . . ."[19]

This is not to say that well-framed national security or military strategy can always accomplish its objectives without combat. Rather, it is to say that a sound strategy (that is, the integrated application of available means) may well yield the desired political result without conflict. However, should conflict occur, sound strategy surely enhances the prospects of achieving desired military and, above all, political outcomes. It is reasonable to interpret Sun Tzu's dictum that "a victorious army wins its victories before seeking battle; an army destined to defeat fights in hope of winning"[20] as meaning that soundly prepared strategy leads to victories. On the other hand, to quote the title of Thomas Mowle's book, *Hope is not a Plan*. The absence of a strategy increases the likelihood of defeat.

FALSE DICHOTOMIES

The Department of Defense (DoD) defines strategy as "A prudent idea or set of ideas for employing the instruments of national power in a synchronized and integrated fashion to achieve theater, national, and/ or multinational objectives."[21] Gabriel Marcella and Stephen Fought find this definition "bureaucratically appealing, politically correct, and relatively useless."[22] For somewhat different reasons, I would agree. First, the DoD definition raises strategy to a transcendent entity — an idea, imbuing it with an ethereal quality that is likely to mystify rather than clarify just what is intended by the term. Second, though I find myself in complete agreement with the DoD's use of the word *integrated*, the use of the word *synchronized* might suggest to some that the available means must be employed in a synchronous or simultaneous fashion. Depending on the circumstances, however, some means may be employed simultaneously while others sequentially — as in a game plan in almost any sport. Finally, the DoD definition wrongly ties strategy to the "instruments of national power," relegating strategy solely to accomplishing "theater, national, and or multinational objectives."[23] Such a definition, of course, accords with what has generally been considered to be *grand strategy* or perhaps *national strategy*, but strips it of its utility as an important tool at every level of human endeavor. For the military, the result has been the establishment of a wall of separation between strategy, supposedly only undertaken by senior political and military officials, and the so-called operational art, undertaken at the theater or campaign level of a conflict.

The U.S. military borrowed the term *operational art* from the Soviets to describe the conceptualization of warfare at the campaign/theater level. Of course operational art isn't an art, or at least not solely art, thus a poor descriptor in the first place for what is intended. The DoD defines operational art as "The application of creative imagination by commanders and staffs—supported by their skill, knowledge, and experience—to design strategies, campaigns, and major operations and organize and employ military forces. Operational art integrates ends, ways, and means across the levels of war."[24] Now that is a lot of bureaucratese to describe thinking strategically at the operational level of warfare, which the DoD subsequently defines as:

> The level of war at which campaigns and major operations are planned, conducted, and sustained to achieve strategic objectives within theaters or other operational areas. Activities at this level link tactics and strategy by establishing operational objectives needed to achieve the strategic objectives, sequencing events to achieve the operational objectives, initiating actions, and applying resources to bring about and sustain these events.[25]

Furthermore, the DoD definition of operational art suggests that designing campaign and major military operations is on an equal footing with designing strategies, rather than products of strategy.

Similarly, the military has established a wall of separation between strategy and tactics, the latter of which it regards as an activity undertaken by lower level officials. As with the word *strategy*, the word *tactics* has a long history, derived from the Greek word *taktika* and its plural *taktikos* or fit for arranging.[26] The Merriam-Webster Dictionary provides as a first definition of tactics: "the science and art of disposing and maneuvering forces in combat," amplifying that with "the art or skill of employing available means to accomplish an end."[27] While the first pays deference to earlier uses related solely to military forces, the latter bit sounds curiously enough like strategy. Regrettably, the DoD defines tactics as "The employment and ordered arrangement of forces in relation to each other,"[28] thus condemning those who operate at the tactical level of warfare to the implementing of procedures and the employment of approved techniques — two synonymies to which one is referred for a better understanding of the term tactics. Thus, we are left with a largely useless definition for the full panoply of tasks undertaken by lower level commanders, particularly given the conditions of modern warfare.

This is, of course, not to deny that commanders at the tactical level often confront problems that are amenable to "engineered" or structured solutions in which repetitive training and the application of approved techniques and procedures significantly increase the prospects for success once militarily engaged with the enemy. However, the modern battlefield seldom mimics classical models, particular in an age of asymmetric warfare. *Ceteris paribus* seldom, if ever, applies as adversaries adjust to American strengths and probe for weakness. Thus, tactical commanders are and will increasingly be required to exercise not just *intuitive* skills based on pattern recognition and procedural responses employing approved techniques, but also reasoned analysis and judgments that bring to bear all available tools at the commander's disposal in order to achieve success.

The point here is that, in reality, success at each level of military activity — strategic, operational, or lower levels — requires that commanders at those levels think strategically, employing in an integrated manner available means to achieve desired ends. Perhaps more importantly, these means should and often do include means beyond those of preparing military forces and engaging in combat. For example, military operations below the campaign or theater level often include working with local leaders and others to provide intelligence and force security (political), cutting of supply routes to adversaries (economic), undertaking local projects to provide safe water or the delivery of food to the local population (economic/psychological), and/or the use of deception to alter the mind set of the adversary (psychological).

Liddell Hart wrote: "In peace we concentrate so much on tactics that we are apt to forget that it is merely the handmaiden of strategy."[79] There is a greater truth in this statement than Liddell Hart had intended. That truth is that those generally accepted tactics (i.e., procedures and employment techniques) are there to serve the game plan of the tactical commander. They do not relieve him of his responsibility to develop a game plan that includes all instruments available to him nor do they relieve those who prepared him for tactical level command of their responsibility to educate and train him in an understanding of and ability to develop strategy at tactical levels.

DELINKING STRATEGY FROM THE WORD STRATEGIC

Perhaps part of the problem we confront in terminology is that as the use of the word strategy was becoming more prominent in the military as well as nonmilitary fields of endeavor, there was a corresponding increase in the use of the word strategic. As a part of Allied strategy for defeating Nazi Germany, the United States and Great Britain undertook "strategic bombing" aimed at crippling Germany's war effort and will to fight. Following the end of World War II, the United States established the Strategic Air Command, whose task it was to deliver a withering blow to the Soviet Union should it attack the United States or its allies. Strategic bombers, submarines, and missiles were defined as those that could deliver weapons over long-ranges, affecting the prospects for the survival of a nation. In juxtaposition, tactical forces were those of lesser reach, which, when employed, had little impact on the survivability of a nation. In business, industry, and education, institutions were charged with developing strategic plans detailing how they would advance their long-term objectives. Unfortunately, the word strategy, particularly through past military usage, too often has come to be linked to its derived cousin, strategic, and has come to signify only what is done at the strategic level of military or business endeavors, rather than being understood in terms of a game plan. One pernicious result is the perception that only national leaders and perhaps senior military officers are the ones who engage in strategy.

POLICY VS. STRATEGY CONUNDRUM

It is common to contend that strategy must follow policy. For example, if it is U.S. policy to support a two-state solution to the Palestinian-Israeli problem, then it is the task of those charged with carrying out U.S. policy in the region to devise a strategy to meet the needs of policy. Similarly, if it is U.S. policy to support democratic movements in foreign countries, then it is the charge of those assigned to implement policy in the various regions and countries of the world to devise strategies to accomplish the task. This is policy as setting objectives.

However, there is another way of looking at policy, that is, policy as a means. For example, the two-state policy set by the U.S. Government is likely to be a part of a broader set of policies with grander objectives. Other policies might include restricting arms flows to Hamas, encouraging human rights and greater democracy in the region, opening a dialogue with Syria in order to find common ground for cooperation, encouraging outside actors to support U.S. efforts in the region, etc. Taken together, these policies thus serve as a means to achieve broader national goals. Such goals might include reducing the probability of conflict, increasing the general welfare of the region's citizenry, reducing likelihood that the region's problems serve as a breeding ground for terrorism, stabilizing the region to ensure the orderly flow of oil from the region and increased stability in world oil markets, and improving global cooperation on vexing problems that threaten the international community. Thus the sum total of such policies, in fact, is (or at least should be) a product of a grander strategy. Under such circumstances, one could properly conclude that policies serve strategy.

Strategy comes first. Then follow policies as the means to accomplish one's strategic design.[30]

This is important to keep in mind, because as one moves from grand or national strategy to policies at multilevels below grand strategy, one must remain aware of the fact that lower level policies are but a means to accomplish national level tasks. Furthermore, as means they remain among a variety of choices governments can make to accomplish desired ends. The danger is always in allowing lower level policies, which serve as means, to become national level objectives. Perhaps this was the case, for example, during the 1960s and 1970s, when, in pursuing the objective of enhancing the security of the nation, the United States engaged in a long war in Vietnam in order to check the worldwide growth of communism. Indeed, for years Vietnam was considered a vital national interest—one worth the shedding of the blood of many young Americans. Following the defeat of South Vietnamese forces by the North, Vietnam ceased to be a vital interest. Had we for many years transformed a means into the end itself, failing to realize, until the administration of President Nixon, that there were other means to enhance the security of the nation?

A more insidious problem in the policy vs. strategy, chicken vs. egg, debate, particularly where military strategy is concerned, is that the very separation of these two terms suggests that there are two clearly identifiable realms of activity. In fact, where national security policy is concerned and the instruments of military power are to be employed (e.g., covert operations, displays of force, deployments, and the wide range of potential employment options), judgments by policymakers must be formed only in close consultation with their military advisers. As Clausewitz noted in his tactical letter to General Muffling: "The task . . . is mainly to prevent policy from demanding things which are *against the nature of war* (italics in original), and out of ignorance of the instruments from committing errors in their use."[31] More importantly, those whose task it is to undertake military activity on behalf of the political goals set by the nations leaders must be well-educated in the strengths and weaknesses of all the instruments of national power, so that they can advise best on what other instruments should be employed and in what manner so as to maximize the usefulness of the military options that might be chosen.

To better illustrate this point, during the troubled times in Central America and the Caribbean in the late 1970s, the U.S. Army Strategic Studies Institute was called upon to undertake a study of the role of the military in that region. When the study was completed, it recommended the United States undertake a number of political/diplomatic and economic initiatives in conjunction with recommended efforts by military personnel. When this study was briefed to a senior military official, that official asked why he needed to know about the political/diplomatic and economic initiatives, since his task was to salute and undertake whatever military tasks were assigned. In response, the briefing team noted that the probability of success of any specific military option hinged on its careful selection from and coordination with the other instruments of national power. Thus, it was the task of senior military leaders to ensure that the nation's political leaders were well aware of the need for a strategy that integrated the instruments into an effective plan to advance U.S. interests in the countries of the region.

ELEMENTS OF THINKING STRATEGICALLY

Some years ago, Kenichi Ohmae in his seminal *The Mind of the Strategist* said: "successful business strategies result not from rigorous analysis but from a particular state of mind."[32] He went on to contend:

> [In] the mind of a strategist, insight and a consequent drive for achievement... fuel a thought process which is basically creative and intuitive rather than rational. Strategists do not reject analysis. Indeed, they can hardly do without it. But they use it only to stimulate the creative processes, to test the ideas that emerge, to work out their strategic implications, or to ensure successful execution of high-potential "wild" ideas that might otherwise not be implemented properly.[33]

One might infer from such a statement that strategists are born, not made. Not so, Ohmae responded, "There are ways in which the mind of the strategist can be reproduced or simulated, by people who may lack a natural talent for strategy . . . there are some specific concepts and approaches that help anyone develop the kind of mentality that comes up with superior strategic ideas."[34]

If Ohmae is correct, what then are these concepts and approaches that, if taught, can help develop good strategists? What then are those universal elements that constitute sound approach to dealing with a problem? What are the concepts that, through practice, will train the mind to think rationally and methodically, yet serve to stimulate the creative processes and thus lead to the development of well-framed game plans, elements that can be applied at all levels of human interaction, whether one is dealing with a crisis, an immediate confrontation, or engaged in long-term planning?

I would suggest seven broad categories of inquiry —
(1) defining the situation, (2) detailing your concerns
and objectives, those of your principal antagonist(s)/
competitor(s), and those of other important players,
(3) identifying and analyzing options that might
be pursued, in terms of such factors as costs, risks,
and probabilities of success, (4) options selection
and alternatives analysis in the light of potential
frictions, (5) reoptimization in light of changing
events, (6) evaluation of the option in terms of its
success in achieving desired results, and finally, (7)
option modification or replacement. The proposed
processes are rational and methodical; yet involve
thinking that is nonlinear as well as multidimensional,
thus stimulating creativity. In examining each of the
elements, I will refer to the development of strategy at
the national level. However, the model can be applied
at all levels of activity.

Defining the Situation.

The first step in developing a sound strategy for
dealing with a problem is to detail the facts of the
situation: what the actual situation is as best can be
known at this point—i.e., the objective, not subjective
reality. In a military environment, this would include
an elaboration of the characteristics of the operating
area, including political, economic, and sociological
factors that may affect operations and a detailing
of enemy, as well as friendly, forces, much akin to
that what is often provided in the Commander's
Estimate of the Situation, though not so cursorily
drawn, as is too often the case. Unconfirmed reports
or speculative information must be set aside for
further investigation—perhaps intelligence tasking.
Statements of values and the ascribing of intentions to

any of the actors should be avoided. Facts are value neutral. At this point, any introduction of values and speculation about the intentions of other players will cloud rather than help clarify the situation. Similarly, interjecting one's concerns and one's own objectives, though one could argue are indeed facts, are steps that should only be taken after the factual situation has been clearly defined. On first blush, this may seem a bit mechanical. However, it provides a necessary clarity essential for the development of effective strategies.

Identifying One's Concerns.

Once the facts of the situation have been detailed, then one should clearly define just what it is that is of concern. What is it that is causing that uneasy state of blended interests, uncertainty, and apprehension? What is it that disturbs or creates angst? Here the trained strategist is disciplined to avoid simply restating the facts, for example, country X has invaded country Y, but rather why should we care? Why should we be concerned? He or she also avoids exaggerating the dangers. Exaggeration of the potential dangers, more often than not, impedes rather than advances the prospects for the emergence of effective strategies, as fear conquers rationality.

Furthermore, the trained strategist will consider not just immediate concerns that emanate directly from the existing problem, but also broader, short-, medium-, and long-term concerns that might be the product of the nonresolution of the current problem. Thus the mind must be trained to wander beyond the confines of the existing issue and the immediate parties to the broader arena of issues among a wider range of parties and interests that might be affected. For example, the testing by North Korea of missiles capable of putting a satellite in orbit, when coupled with their continued development and acquisition of nuclear weapons, not only raises concerns about stability on the Korean peninsula, but also a wide variety of concerns ranging from the future of stability, arms races, and the proliferation of nuclear weapons in Asia to the future dangers such developments might pose for America's security.

Where a developing situation raises multiple concerns, as is most often the case, concerns then must be prioritized. For example, if a country such as Iran is seeking to acquire nuclear technology ostensibly for the production of nuclear energy, the U.S. President may be concerned that those materials might be used in the production of nuclear weapons. He also might be concerned that such weapons, if developed, might upset the balance of power in the region in which the country is located, undermine U.S. interests and those of friends and allies, and result in a further breakdown in efforts to limit the proliferation of nuclear weapons and the spread of such technologies to terrorist groups and others bent on doing harm. Furthermore, the President might well be concerned that such weapons could be used against one or more friendly countries in the region, or might result in a preemptive or preventive attack by one of the threatened countries and subsequent regional conflagration, eventually forcing the United States to take military action with its attendant loss of innocent lives and potential regional and global political and economic implications. Additionally, he might be concerned that any failure to act on his part may be perceived by Iran as well as others, including some in the United States, as weakness. Countries in the region might start paying deference to Iran, and/or other countries reliant on the security provided by the United States might lose confidence in those guarantees. All of such concerns are not of equal weight. Prioritizing concerns before making recommendations to the President enables the strategist to analyze and evaluate options for dealing with the problem in terms of their ability to address, if not all concerns, the most critical ones.

Identifying One's Objectives.

Once concerns have been identified and prioritized, it is then time to specify one's short-, medium-, and long-term objectives for the country, region, and worldwide objectives. A number of objectives may be long-standing in nature or an outgrowth of current events or both. For example, in the Iranian example noted above, an objective of preserving or improving regional stability not only would be a reflection of long-standing American policy, but also the result of concerns raised by the emerging crisis.

However, objectives should also be viewed in an expansive context. Sound strategic thinking at the national level demands that seemingly unrelated regional and global objectives also be understood and delineated. In today's globalized world, crises and their solutions seldom exist in isolation. Actions in one part of the world often beget actions, even if not equal and opposite, in other parts of the world. Thus, it is imperative that strategists have a well-rounded understanding of the broader policy objectives before undertaking analyses of potential options for dealing with given situations.

Though the contention that the Chinese pictograph for crisis is made up of two characters, one standing

29

for opportunity, the other for danger, is a matter of dispute, history is replete with examples of opportunities derived from danger. Peoples have been mobilized, decisions made, and energies expended that would not otherwise have occurred in the absence of a crisis and the dangers it entailed. Thus nearly every crisis affords the opportunity to advance or, depending on the policy options chosen, endanger the successful accomplishment of broader objectives. Thus, for example, the United States might have such broader political objectives as improving relations with Russia and China, forging a just peace in the Middle East, and further advancing cooperation with and among our European allies. A clear understanding of such broader objectives would permit strategists seeking solutions, say to the Iranian dilemma noted above, to evaluate policy options in terms of their impact on such broader objectives.

Perhaps more importantly, where policy objectives are unclear, poorly articulated, and/or in conflict with one another, the strategist must be a visionary, identifying the road ahead, clarifying objectives, and engaging in carefully articulated discussions with those responsible for setting the broader national or military objectives. In simple terms, to travel the correct road, you need to know where you are going. For example, at the end of World War II, President Truman ultimately rejected the plan of Secretary of the Treasury Henry Morgenthau, Jr., which, among other things, would have divided Germany, allowed for the annexation of parts of Germany by its neighbors, and reduced Germany to an agrarian state. President Truman opted instead for a united Germany and a policy of economic reconstruction. By 1951 the Truman administration also had spent about $12.4 billion under the Marshall

Plan to assist Europeans in their economic recovery. Such efforts gave both the Germans and others hope of a brighter future, which has resulted in a historically unprecedented era of peace and cooperation in all of Western Europe.

No such vision accompanied U.S. assistance to Afghanistan following the 1979 Soviet invasion. When the last Soviet troops were withdrawn from Afghanistan on February 15, 1989, following nearly a decade of war, the United States abandoned Afghanistan. Thus Afghanistan was left to deal with its own problems of political and economic stability and the explosion of Taliban influence and subsequent human rights violations. Today the United States continues to suffer the consequences of this lack of foresight.

Like concerns, objectives also should be prioritized. Failure to do so may ultimately lead to choosing options for dealing with a situation that, while they successfully resolve the current problem, place in jeopardy higher priority regional and global goals. For example, some have argued that, while it may have been laudable for the United States to remove the brutal dictator Saddam Hussein, the invasion of Iraq became the poster child for recruiting terrorists around the world, thus undermining a major post 9/11 objective of American foreign policy.

Identifying the Objectives and Concerns of Others.

Understanding the objectives and concerns of the principal antagonist(s), as well as other principal players, is of paramount importance in devising any game plan. Here informed speculation can play a significant role. One can seldom know with a high degree of certainty the objectives and concerns of others, particularly nation-states. Indeed, actions often may reflect bureaucratic, institutional, or political factors that are not easily accounted for in a simple rational actor model of behavior. Thus, in-depth knowledge of such factors as the country's history, culture, past actions, and those bureaucratic, institutional, and political factors that might affect the country's decisionmaking processes is required. Former Secretary of Defense Robert McNamara correctly identified one of the major reasons for our failed strategy in Vietnam, noting that our judgments of friend and foe, alike, reflected our profound ignorance of the history, culture, and politics of the people in the area, and the personalities and habits of their leaders.[35]

A trained strategist does not necessarily require such knowledge, though it would enhance his ability to undertake informed speculation. However, in the absence of such skills, the strategists must surround themselves with those who do, and be trained to ask the right questions.

The question, of course, that always arises is: What if the adversary behaves irrationally? Without disputing the fact that individuals and groups may act irrationally, their actions, from their point of view, seldom, if ever, are perceived as irrational. Thus, an understanding of what motivates the behavior of leaders, what they seek, what they fear, what may drive them to make decisions that from our perspective may seem irrational, is essential in the formulation of sound political and military strategies.

The absence of an understanding of such factors may have led to a profound strategic failure that culminated in the 2003 Iraq War. The White House continued to believe, despite significant if not overwhelming evidence to the contrary, that Saddam Hussein had weapons of mass destruction (WMD). In August 1995 General Hussein Kamal, the defecting son-in-law of Saddam Hussein, had reported to senior United Nations (UN) officials: "All weapons—biological, chemical, missile, nuclear were destroyed."[36] UN inspectors, despite having the best available intelligence from the United States and other countries, were unable to discover any WMD. Other evidence suggesting that Saddam Hussein had continued or renewed his efforts to acquire WMD rested on thin reeds.[37] One can imagine that from the White House perspective, given the circumstances of an impending attack by the United States and other allied forces, it simply would have been irrational for Saddam Hussein not to take all steps necessary to assure the United States that Iraq did not possess such weapons. But, according to the post-invasion Duelfer Report which confirmed that no WMD could be found, Saddam, greatly weakened following the war with Iran which ended in 1988 and the Gulf War of 1991 and concerned about his enemies, did not want to appear weak and therefore was deceiving the world about the presence of WMD.[38] The result: a long war that has cost the United States dearly in lives, treasure, and reputation, and more than likely added fuel to the flames of terrorism.

Options Identification and Analysis.

The next step in the process is to identify potential options that might exist that can advance one's objectives, while allaying or limiting one's concerns and to analyze the costs and risks that each option or group of options entails. At the level of grand/ national strategy, options usually include one or more instruments of national strategy, which are the multifaceted means that are to be used to accomplish desired ends. Such instruments usually fall into such categories as political/diplomatic, informational, economic, psychological, and military. Options may include the use of two or more instruments simultaneously or sequentially or both or primary reliance on a single instrument.

For example, during the Gulf crisis and war of 1990-1991, the administration of George H. W. Bush, determined that Saddam Hussein's occupation and annexation of Kuwait should not be allowed to stand, reached into its tool bag of implements, and selected a number of political/diplomatic, economic, and military instruments. Among those instruments used, diplomacy initially was employed primarily to garner support for the removal of Saddam's forces from Kuwait. Economic sanctions, though often imperfect in effect, were employed to demonstrate to Saddam and others the severity of the situation and perhaps as a necessary step in the process of getting later approval for the employment of force. Since publics and nations often expect the use of all means short of war before agreeing to the use of force, the economic instrument may play both an economic and a psychological role. Later the economic instrument, including promises of aid, debt forgiveness, and direct payments, was used in conjunction with the diplomatic instrument to encourage support by other nations for military efforts to expel Iraqi forces from Kuwait. Additionally, significant numbers of ground, air, and naval forces were deployed to the region to prevent Saddam's ambitions from extending to the Kingdom of Saudi Arabia, to serve as a warning that failure to comply with UN resolutions calling for a withdrawal of forces might result in war, and later to force Iraqi withdrawal.

Shortly following the Iraqi invasion of Kuwait, the psychological instrument also was employed. To those concerned about what kind of order the post-Cold War world would involve, Bush linked the success of a "new world order, a world where the rule of law, not the law of the jungle, governs the conduct of nations,"[39] with the international community's response to the invasion of Kuwait. To those appalled by such overt aggression, the Bush administration raised the specter of another Hitler, this time in the Middle East. To those concerned about the cost of living and future economic progress, the administration linked failure to firmly confront Iraq's invasion of Kuwait with high oil prices, and declining economies. The psychological instrument proved helpful in securing the support of the American public and a favorable Congressional vote to authorize the use of military forces to end Iraqi occupation of Kuwait.

This was a nonlinear, multidimensional, simultaneous, and sequential use of multiple instruments of national power to achieve national objectives — in short, a well-framed strategy. On August 2, 1990, the very day Iraq invaded Kuwait, all five permanent members and nine of the other 10 members (Yemen did not vote) of the UN Security Council voted in favor of UN Security Council Resolution 660, condemning the Iraqi invasion of Kuwait and demanding the withdrawal of Iraqi troops. Four days later, 13 members of the UN Security Council voted in favor of Resolution 661, placing economic sanctions on Iraq (Cuba and Yemen abstained). On November 29, 1990, 12 members of the UN Security Council voted in favor of Resolution 678 (Cuba and Yemen voted against and China abstained), which gave Iraq until January 15, 1991, to withdraw from Kuwait and authorized "all necessary means to uphold and implement Resolution 660,"[40] a diplomatic formulation authorizing the use of force. On January 12, the U.S. Congress authorized the use of U.S. military forces. On January 17, the air war began. On February 24, allied ground forces began their attack. Thirty-four countries lent their support. Within about 100 hours of the initial ground assault by allied forces, the world's fourth-largest army was defeated.

On the other hand, there are times when a single instrument of power has been the primary tool in attempts to advance American policies. This, for example, has been for the most part the case in U.S. attempts to achieve a just settlement in the Middle East, where it has often relied primarily on diplomacy with an occasional suggestion of the use of the economic instrument in efforts to cajole parties in the Middle East to the American point of view.

Understanding the objectives and concerns of the adversaries or potential adversaries and other principal players — what they seek, what they value, and what they fear — is a major ingredient in identifying how their behavior can be influenced. Thus, the option(s) ultimately selected not only should promise to allay U.S. concerns and advance U.S. objectives within bearable costs and risks, but also should be formulated in such a way that failure on the part of the other actors to adopt behavior in line with U.S. preferences would lead to an increase in their concerns and a reduction in the possibility that they would achieve their objectives. Ideally, adoption of the U.S. preferred option(s) also would allay some, if not all, of their concerns and advance some of their objectives. In other words, at the national level wise policies seek to create the perception, if not the reality, of a win-win scenario. This, of course, was the strategy pursued by the United

States and the Soviet Union as they entered into arms control negotiations begun in Helsinki, Finland, in 1969.

On the battlefield, of course, this non-zero sum, win-win approach often fails the test of reason, since the object of combat is defeat of the enemy. Yet the basic principles remain, where available means are used to alter and direct the behavior of an adversary, perhaps luring him to actions that favor his defeat. The use of deception as a tool to affect the psychology and thus decisionmaking of Hitler prior to the invasion at Normandy is a prime example.

The options development phase of strategy is the phase that demands the greatest degree of creativity. Too often this is the weakest point. Options are frequently too narrowly drawn. Choices are sometimes framed in terms of three options — one at one extreme, the other at the other extreme, and one somewhere between — that all reasonable decisionmakers are expected to elect. Or perhaps choices are framed even more narrowly — concede/surrender or fight. All too often, options are the product of linear thinking. Typical of a linear approach is a formulation and analysis of options that focus solely on solutions to the existing problem. Thus, linear thinking often fails to consider an option's medium- and long-term impact on the objectives and concerns of other players, as well as on the objectives and concerns that seemingly stand quite apart from the contemporary problem, perhaps relating to issues and countries not directly affected or involved in the current situation. In short, the strategist must have an understanding of the entire strategic environment at his or her level of activity if an effective strategy is to be devised.[41]

The well-trained strategist also understands that, "as with other aspects of life, there may be problems for which there are no immediate solutions. . . . At times, we may have to live with an imperfect, untidy world."[42] On the other hand, good strategy is not risk free. Seeking risk free options is a common prescription for inaction or failure.

Options Selection and the Frictions and Fog of Events.

This is the final stage of the initial process of strategy building. Each multifaceted option, having been rationally examined in terms of its costs and risks, is exposed to the scrutiny of the strategist in terms of its probability of allaying concerns and advancing objectives. It is at this stage that intuition can play a significant role. Intuition is not a guess. It is the "power or faculty of attaining direct knowledge or cognition without evident rational thought and inference."[43] It is a quick and ready insight, that immediate understanding that comes from previous knowledge and experience. Thus a successful strategist is likely to be one who has a sound understanding of the players (at the national level—other nations or nonstate actors; in military situations—of opposing forces and their leadership), a well-rounded knowledge of the strengths and weaknesses of the various instruments at his or her disposal, and enough experience to know that seldom if ever do things go according to plan.

Of course this is what Clausewitz labeled "friction." To paraphrase Clausewitz, everything may look simple, the knowledge required may seem to be at hand, and the strategic options may seem obvious. However, once the clash of wills is engaged, stuff happens. Or as Moltke put it: "No plan of operations survives the first collision with the main body of the enemy."[44] However, it would be wrong to conclude as

Moltke that under such circumstances strategy is little more than a "system of expedients."[45] Rather it is to underscore and broaden the context of a view of war held by Marshall Maurice de Saxe: ". . . it is possible to make war without trusting anything to accident."[46] Factoring in the potential for frictions to arise and for situational changes that may affect the game plan is a part of thinking strategically. Thus it is at this stage that the strategist must be trained to ask the "What if" question. What if things do not go according to plan? What additional alternatives remain? Again, not unlike sports, all other things being relatively equal, success comes to those who are best able to respond flexibly, to plan for and pursue alternative courses of actions should their preferred approach fail to succeed. Indeed, to paraphrase a cardinal principle of French General Pierre-Joseph de Bourcet, who was infected by thinking similar to that of de Saxe: a game plan should have several branches.

One should study the possible courses of action in the light of the obstacles to be overcome, of the inconveniences or advantages that will result from the success of each branch, and, after taking account of the more likely objections, decide on the part which can lead to the greatest advantages, while employing diversions and all else that one can do to mislead the enemy and make him imagine that the main effort is coming at some other part.[47]

Failure to ask the "What if" question and plan for alternative approaches may well have been the single most significant factor that has resulted in a long-term, costly engagement in Iraq. Though warned beforehand that large numbers of forces would be required to keep the peace in Iraq following any successful invasion, President George W. Bush chose the comfort of rosy predictions rather than ask such critical questions as: What if chaos ensued and things went south? What might be the resulting implications for the American game plan? What steps should be taken ahead of time to either preclude chaos or bring quick order to Iraq to prevent an ensuing breakdown in the social order that surely would be costly in terms of additional lives lost and might threaten the very success of the objectives sought by the invasion in the first place?

FOLLOW-ON ACTIVITIES

Sound strategies never end with the implementation of the selected option. Constant vigilance is demanded with an eye toward ever evolving situations. Thus any selection of means will require a re-optimization in light of changing events and then evaluation in terms of the success in alleviating concerns and achieving objectives relative to the current situation, as well as other short-, medium-, and long-term concerns and objectives. Modifications will be made, which in turn will require further evaluation, in a continuing process, which may see major alterations to the original plan. In this regard, strategists must retain a flexibility of mind until such time as the designated objectives are achieved.

CONCLUSIONS

Strategy can best be understood as the integrated application of available means to achieve desired ends. At the national level such means usually include a combination of political/diplomatic, informational, economic, psychological, and military instruments. However, the need to think strategically permeates all levels of decisionmaking. False dichotomies, which suggest that strategy is what is undertaken at higher levels of government or the military and tactics is what lower levels undertake, are not only misleading, but also counterproductive. Individuals must be trained to think strategically at all levels. Only then can they employ the means at their disposal in ways that maximize the probability of achieving success.

Also misleading is the artificial separation of policy and strategy. Policies understood as objectives cannot succeed without a corresponding strategy for achievement. Likewise, the aggregation of policies, understood as means when well thought through and well-integrated, constitute a strategy.

The primary task with which we are confronted is to educate and train individuals to think strategically at all levels of endeavor. This chapter has identified those elements that, if practiced iteratively, will help train the mind to think methodically, rationally, and creatively, that is, to think strategically. There are those who come by such methods naturally but, as with good artists and scientists, most are educated to their profession. As we look to the future, the need for strategic thinking and sound strategists will be at a premium. We must therefore develop a solid cohort of those who can do so, whether they are dealing with a crisis, handling an immediate confrontation, or engaged in long-term planning.

ENDNOTES - CHAPTER 2

1. Harry G. Summers, Jr., *On Strategy: The Vietnam War in Context*, Carlisle, PA: Strategic Studies Institute, U.S. Army War College, 1981, p. 1.

2. "Overview," The Human Security Report 2005: War and Peace in the 21st Century, New York: Oxford University Press, 2005, p. 1.

3. Francis Fukuyama, "The End of History?" *The National Interest*, Summer 1989, p. 18.

4. Moisei Ostrogorski, *Democracy and the Organization of Political Parties*, Vol. 2, Frederick Clarke, trans., New York: MacMillan Co., 1902, p. 579.

5. Alexis de Tocqueville, *Democracy in America*, Vol. 1, The Henry Reeve Text as revised by Francis Bowen and further edited by Phillips Bradley, New York: Random House Vintage Books, 1990, p. 235.

6. Clyde Kluckholm, who died in 1960, had an enduring influence on the anthropological study of culture.

7. Walter W. Skeat, Rev., *A Concise Etymological Dictionary of the English Language*, New York: G. P. Putnam's Sons, 1980, p. 522.

8. Carl von Clausewitz, *On War*, Michael Howard and Peter Paret, eds. and trans., Princeton: Princeton University Press, 1976, p. 177.

9. Quoted in B. H. Liddell Hart, *Strategy*, London, United Kingdom: Faber & Faber Ltd, 1967, p. 320.

10. Liddell Hart was well aware of the greater subtleties of Clausewitz's approach to strategy. He noted that Clausewitz admitted that "the object of a combat is not always the destruction of the enemy's forces," and "its object can often be attained as well without the combat taking place at all." His criticism of Clausewitz was that everyone would catch such ringing phrases as "We have only one means in war—the battle," "The combat is the single activity in war," "We may reduce every military activity in the province of strategy to the unit of single combats," and "let us not hear of generals who conquer without bloodshed." See B. H. Liddell Hart, *The Ghost of Napoleon*, New Haven, CT: Yale University Press, 1935, pp. 124-126.

11. Liddell Hart, *Strategy*, p. 321.

12. Clausewitz, *On War*, p. 87.

13. Merriam-Webster Online, available from *www.merriam-webster.com/dictionary/strategy*.

14. *Ibid.*

15. Clausewitz, *On War*, p. 149.

16. Samuel B. Griffith, *Sun Tzu: The Art of War*, New York: Oxford University Press, 1963, p. 77.

17. *Ibid.*, p. 79.

18. *Ibid.*, p. 77.

19. B. H. Liddell Hart, "Thoughts on Philosophy, Politics and Military Matters," June 7, 1932, Liddell Hart Papers II/1932/20, quoted by Christopher Bassford, *Clausewitz in English: The Reception of Clausewitz in Britain and America, 1815-1945*, New York: Oxford University Press, 1994.

20. Griffith, *Sun Tzu: The Art of War*, p. 87.

21. See *Joint Publication (JP) 1-02, Department of Defense Dictionary of Military and Associated Terms*, Washington, DC: Joint Chiefs of Staff, April 12, 2001, as amended through October 17, 2008, p. 525.

22. Gabriel Marcella and Stephen O. Fought, "Teaching Strategy in the 21st Century," *Joint Force Quarterly*, Issue 52, 4th Quarter 2009, p. 60, note 2.

23. *JP 1-02.* p. 525.

24. *Ibid.*, available from *www.dtic.mil/doctrine/dod_dictionary/*.

25. *Ibid.*, p. 399.

26. Skeat, *Etymological Dictionary*, p. 539.

27. Merriam-Webster Online, available from *www.merriam-webster.com/dictionary/tactics*.

28. *JP 1-02*, available from *www.dtic.mil/doctrine/dod_dictionary/*.

29. B. H. Liddell Hart, *Thoughts on War*, London, UK: Faber and Faber, 1944, p. 48.

30. On the other hand, if one understands policy in terms of one of the definitions provided by Merriam-Webster, as "a high level overall plan embracing the general goals and acceptable procedures, especially of a government body", then at the national level policy is strategy by another name. Unfortunately, however the term policy seldom seems to reflect the degree of integration of available means to achieve desired ends that is conveyed by the term strategy.

31. Daniel J. Hughes, ed., *Moltke on the Art of War: Selected Writings*, New York: Ballantine Books, 1993, p. 36.

32. Kenichi Ohmae, *The Mind of the Strategist: The Japanese Art of Business*, New York: McGraw-Hill, Inc., 1982, p. 4.

33. *Ibid.*

34. *Ibid.*, p. 5.

35. Robert S. McNamara, *In Retrospect: The Tragedy and Lessons of Vietnam*, New York: Times Books. 1995, p. 322.

36. On August 22, 1995, the Executive Chairman of the Special Commission, along with a member of the International Atomic Energy Agency and another member of the Special Commission met with General Hussein Kamal in Amman, Jordan. For a transcript of that meeting, see "Note for File," UNSCOM/IAEA Sensitive, available from *www.fair.org/press-releases/kamel.pdf*. For the specific quote, see page 13.

37. For example, see Robert Kennedy, *Of Knowledge and Power: the Complexities of National Intelligence*, Westport, CT: Praeger Security International, 2008, pp. 84-85, 95-100, 149-157.

38. See "Regime Strategic Intent, "*Comprehensive Report of the Special Advisor to the DCI on Iraq's WMD*, September 30, 2004, available from *https://www.cia.gov/library/reports/general-reports-1/iraq_wmd_2004/index.html*.

39. George H. W. Bush, *Address to the Nation Announcing Allied Military Action in the Persian Gulf*, January 16, 1991, George Bush Presidential Library and Museum, Public Papers 1991, available from *bushlibrary.tamu.edu/research/public_papers. php?id=2625&year=1991&month=01*.

40. United Nations Security Council Resolution 678, available from *daccess-dds-ny.org/doc/RESOLUTION/GEN/NR0/575/28/IMG/ NR057528.pdf?Open Element*.

41. See Richard Yarger's sixth premise. Richard Yarger, *Towards A Theory of Strategy: Art Lykke and the Army War College Strategy Model*, available from *dde.carlisle.army.mil/authors/stratpap. htm*.

42. McNamara, *In Retrospect*, p. 323.

43. Merriam-Webster Online, available from *www.merriam-webster,com/dictionary/intuition*.

44. Hughes, *Moltke*, p. viii.

45. Hughes, *Moltke*, pp. ix, 47.

46. Liddell Hart, *Ghost of Napoleon*, p. 30. Marshall de Saxe (1696-1750), considered by some as the greatest European general and military intellectual in the early to mid-1700s, was a French military commander whose *Mes reveries* (My Reflections), later published in English as *Reveries upon the Art of War*, framed the efforts of many military officers who followed. See *Ibid.*, pp. 30-50.

47. Liddell Hart quoting Bourcet. See Liddell Hart, *Ghost of Napoleon*, p. 56.

CHAPTER 3

THE STUDY OF STRATEGY:
A CIVILIAN ACADEMIC PERSPECTIVE

Robert C. Gray

INTRODUCTION

There are at least four reasons to study a nation's strategy for preserving its security. First, strategy is a topic of great historical importance. The strategies devised by politicians and military leaders help determine the fate of nations. Second, many people find strategy to be interesting. At Borders and Barnes & Noble bookstores, the section on war is one of the largest, and, although many of those books are memoirs or descriptions of operations, many contain discussions of strategy as well. Third, classical writings on strategy can be applied to other fields. Sun Tzu and Clausewitz, for example, have been given new life as consultants in business management. Finally, the systematic study of strategy can increase the probability of success in designing new strategies that will enable a nation to attain its security goals in an ever changing world. Indeed, leaders have an obligation to employ strategy so that power is used in ways that are ethical and politically effective. Without strategy, war could degenerate into mindless and ruthless destruction.

While it is edifying to learn about strategic history and while businesses and other organizations may profit from thinking strategically, the most vital reason for understanding strategy is to safeguard the security of one's nation. It is this instrumental reason for studying strategy that is the domain of the staff colleges and war colleges that educate military officers (and, in the case of the war colleges, some civilian government managers). Because of the importance of educating military leaders to think strategically, it is necessary from time to time to assess the teaching of strategy. The mismatch between strategy and reality in the first years of the Iraq War suggests that this is a good time for such an inquiry.

For a democracy, it is also of vital importance for civilians, especially elected officials and staff but also members of the attentive public, to have an understanding of strategy. The responsibility for providing this education lies with our nation's colleges and universities. The purpose of this chapter is to discuss strategy and some ways of teaching it from the perspective of a professor who teaches about strategy at a liberal arts college.

DEFINITIONS AND LEVELS OF STRATEGY

Despite the differences between the worlds of civilian education and professional military education (PME), some challenges are the same. We need to define strategy and its relation to policy. There are two broad views of this. One sees strategy as the foundation of policy. Terry L. Deibel, for example, views strategy "... as an input to the [policy process], a guiding blueprint whose role is to direct policy, to determine what the government says and does."[1]

The second and more common view reverses this and depicts policy as directing strategy. Colin Gray, for example, defines strategy as "... the use that is made of force and the threat of force for the ends of policy."[2] Gabriel Marcella and Stephen Fought define strategy in a similar way as "the art of applying power to achieve objectives, within the limits imposed by policy."[3] This second view has a long tradition dating to Clausewitz and is certainly the proper concept for military strategy. That Deibel has a different view may reflect the fact that he focuses on a strategy for foreign affairs, a much broader activity.

Deibel suggests the following levels of strategy:

National Strategy (domestic as well as foreign),
  Foreign Affairs Strategy (all foreign policy related),
    National Security Strategy (foreign, but security interest only),
      Grand Strategy (broadest conduct of war with all tools),
        Military Strategy (use of military instrument only).[4]

As he notes, there is little agreement on the distinctions between what he calls foreign affairs strategy, national security strategy, and grand strategy. The boundaries between military strategy and the levels above it are not fixed. This ambiguity raises a host of questions. Do the diplomatic, informational, military, and economic (DIME) categories of power cover all relevant variables? At what point does grand strategy become indistinguishable from foreign policy? Should we limit our focus to issues related to national security, or should we include in the teaching of strategy all aspects of foreign policy and some dimensions of domestic policy?

The traditional approach in both the security studies subfield of political science and in the curriculum of the war colleges has been to focus on security-related issues. Since it was first suggested in 1943, Edward Mead Earle's definition of grand strategy has been useful: "[t]he highest type of strategy . . . which so integrates the policies and armaments of the nation that the resort to war is either rendered unnecessary or is undertaken with the maximum chance of victory."[5] In terms of Deibel's levels of strategy, the focus of inquiry at the war colleges is most likely to be the bottom three categories: national security strategy, grand strategy, and military strategy.

One difference between military strategy and grand strategy is that the former is easier to design because it is a more clearly defined domain. In addition, as General André Beaufre reminded us in the 1960s, the military already has the concept of an overall strategy: "to coordinate action on land, in the air and on the sea."[6] The other elements of grand strategy (e.g., politics, economics, and diplomacy) have less of a tradition of coordinated action, though in recent years the Department of State under the leadership of Secretary of State Colin Powell began publishing long-term strategic plans. While the attention paid to grand strategy in the last 40 years has presumably improved our ability to coordinate strategy in nonmilitary fields, grand strategy seems destined to remain underdeveloped in comparison with its military counterpart.

THE STUDY OF STRATEGY AND WAR IN CIVILIAN COLLEGES AND UNIVERSITIES

The Decline in the Academic Study of Military History.

One building block for the study of strategy is the history of its use. Knowledge of military history must be part of the core knowledge of anyone who wants to deal seriously with strategy. In most civilian colleges and universities, however, there are few opportunities to study military history. According to a recent article, only 12 of the 150 or so universities with Ph.D. programs in history offer substantial work in military affairs, and only a few of these are the top departments in the field.[7]

This author surveyed the online catalogs of the history departments in the top 15 liberal arts colleges in the 2008-09 *U.S. News and World Report* rankings.[8] Several of the colleges offer courses on the Crusades and on the American Civil War that appear to deal substantially with causes, consequences, and social implications. One college offers a course on diplomatic history, another offers a course on strategy and diplomacy, and a third offers a course on Vietnam and a course on World War I. Six colleges offer no courses at all that appear to deal with military issues. No college offers a survey course on military history. [9]

The decline in the study of military history can be traced in part to the impact of the Vietnam War, which tarnished any interest in military affairs in much of the academic world. New topics and theoretical perspectives rose to prominence. In the decades after Vietnam, the attention of professional historians shifted from political and military issues to social, cultural, ethnic, racial, and gender history. There are still courses on war, but the emphasis is on its social impact rather than issues of strategy and the use of force. Although the new topics and conceptual approaches have advanced knowledge in important ways, this has come, in a world of scarce resources, at the expense of courses that deal with diplomatic history, strategy, and war.

Despite the negative trend in the availability of courses on military history, however, there has been some outstanding scholarly work in the last 25 years on both grand and military strategy. John Lewis Gaddis' examination of Cold War national security policy, *Strategies of Containment*, is a case in point, as are two edited collections: Peter Paret's 1986 update of *Makers of Modern Strategy*; and the 1994 volume, *The Making of Strategy*.[10]

The Rise of Security Studies in Political Science.

Although students find it difficult to study strategy and war in history departments, they can still do so in many political science departments. Strategic studies, more often called security studies now, emerged in the 1950s. The advent of nuclear weapons, coupled with the emergence of the Soviet-American Cold War rivalry, challenged many assumptions of existing military thought. The principal analysts who examined the implications of nuclear weapons were civilian. Prior to this period, they would have had little to contribute to military analysis because civilians without substantial military experience were in no position to analyze the use of the tank or field artillery. By contrast, no one had experience with nuclear war, and that put civilians on even ground with the military.

Much of the early work was done in think tanks, particularly the RAND Corporation. The economists, political scientists, game theorists, and mathematicians there laid the foundations for what would become strategic studies. It is not possible in the space available here to describe the evolution of the field in the intervening decades. There are useful summaries by Colin Gray, Stephen Walt, Richard Betts, and Edward Kolodziej that deal with the history, theories, methods, and disputes of the discipline.[11]

Security studies is interdisciplinary in nature but is most often housed in the international relations subfield of political science. There has been a substantial amount of work on strategy. Dan Reiter and Curtis Meek summarized the political science literature on strategy up to 1999.[12] They divide strategy into three types: maneuver, attrition, and punishment. Treating strategy as the dependent variable, they aim to identify factors that lead states to choose among these different types. An alternative approach that views strategy as the dependent variable (used, for example, by Barry Posen and Jack Snyder) categorizes strategy as offensive or defensive and attempts to explain why nations choose one or the other. Still others, such as John Mearsheimer, treat strategy as the independent variable. These scholars are interested in the impact of strategy on the likelihood of war and on war outcomes.[13]

Two additional approaches should be mentioned. The use of rational choice theory has become an increasingly influential part of political science, and international relations is no exception. Those who apply economic and game theory use formal models to explore the political world. This work is not very helpful in the study or teaching of grand or military strategy. As Stephen Walt concluded, ". . . recent formal work has relatively little to say about contemporary security issues."[14]

The dominant approach in security studies, Realism, explains state behavior in terms of material interests. Constructivism seeks to provide an alternative explanation of state behavior in terms of ideas, culture, and identity. Alexander Wendt has summarized two key principles of constructivism. The first is "that the structures of human association are determined primarily by shared ideas rather than material forces."[15] The second is "that the identities and interests of purposive actors are constructed by these shared ideas rather than given by nature."[16] Often relying on narratives rather than formal models, constructivism is the methodological opposite of rational choice theory. Taken to extremes, constructivism has little to say to teachers and practitioners of strategy. Paying attention to culture and ideas can, however, be useful. Elizabeth Kier's work on British and French military strategy is a good example of how a moderate constructivist approach can enrich our understanding of strategic issues.[17]

The security studies research that is most relevant to the teaching of strategy has a clear focus on policy. The leading journal in the field, *International Security*, is the most prominent venue for this type of work. As Steven E. Miller, editor-in-chief, wrote on the occasion of the journal's 25th anniversary, ". . . the ultimate goal of our collective endeavor in this field is applied learning. Knowledge and understanding should be sought not for their own sake but to improve the ability of the human race to address security challenges in the safest and most effective possible manner."[18] The applied research that characterizes this branch of security studies focused during the Cold War on topics such as nuclear strategy, Soviet defense spending, and the conventional force balance in Europe. Illustrative topics in recent years include intervention in civil wars, possible conflict stemming from scarcity of resources, and the dynamics of terrorist groups.

In sum, political scientists who specialize in international relations and security studies use diverse theories and methods to examine the use of force. The scholars who engage in applied research provide a particularly relevant body of literature for those interested in the formulation and implementation of strategy in today's world.

WHY COURSES ON STRATEGY AND WAR MATTER

The most striking trends in the academic study of strategy are the decline in the teaching of military history and the rise of security studies in political science. Although the latter development means that strategic issues are being addressed in the academic world, the virtual disappearance of military strategy from history departments still leaves a void.

The availability at the undergraduate level of courses on strategy and war is important for several reasons. First, such courses address complex events that are particularly appropriate for the liberal arts mission of developing analytical thinking. The disagreements over Allied strategy in World War II, for example, are a fascinating window into how rival American and British conceptions of national interest were fashioned into successful military strategies.[19] Simply put, studying strategy is an excellent way to develop critical thinking skills while learning about some of the most important events in the history of the world.

Second, there are students in civilian colleges and universities who want to become military officers. Although they will encounter some military history in Reserve Officer Training Corps (ROTC) programs, their understanding of the important issues in strategy and war would be enhanced by exposure to professional historians in a civilian setting. Peter Feaver and Richard Kohn have described the substantial attitudinal differences between civilians and those in the military.[20] It is not in the nation's interest for this gap to continue, let alone to widen. Indeed, Kohn himself defines good civil-military relations as essential to the process for making strategy in a democracy. We should have an officer corps that is broadly educated at diverse institutions, but that is only possible if students interested in the military feel welcome on civilian campuses.[21]

Having military officers do graduate work on civilian campuses fosters healthy civil-military relations at the broadest level. Because of the small percentage of Americans who join the military, the only contact many civilian graduate students will have with the military is with officers who are in their classes. For military officers, it can be helpful to learn perspectives and approaches from the civilian world. Although this is not a panacea for civil-military relations, anything that helps reduce the attitudinal differences between those in the military and civilians is welcome.

Third, there are students who have neither prior interest in the military nor any desire to wear a uniform. They will, however, be citizens whose government may, from time to time, contemplate the use of force. Some of these students may serve in government — in the Executive Branch, in Congress, or in the courts — where their perspectives on the military will influence their policy or legal judgments in decades to come. Whether a student becomes a government official or is an attentive citizen, the perspectives afforded by college courses on military history will be useful.[22]

Students can learn something about war and strategy in political science courses on international relations, foreign policy, and security policy, but these courses typically cover history only when it is necessary to understand the theoretical and other matters that are central to the field. Colleges today provide less coverage of war and strategy than they once did, and this has consequences for public perceptions. As Peter Paret observed in 1986:

> In the training of historians and the teaching of history, particularly in the United States, war has never been a favorite subject. One result has been to leave far too much scope for a popular, essentially romantic literature on war, which explains nothing, but crudely responds to the fascination that war past and present exerts on our imagination and on our wish to understand.[23]

War is too important to be left to purely popular histories that portray war in simplistic or romantic terms.

APPROACHES TO TEACHING STRATEGY

Strategy at All Levels.

There is a growing recognition that strategic thinking must be taught not just to the few who will be formulating strategy, but to those who will be executing it as well. A Soviet General, Aleksandr Svechin, noted that confining the study of strategy to the top commanders ". . . destroys mutual understanding between staffs and line units. Strategy should not become a kind of Latin which separates the believers and the nonbelievers."[24] To be sure, General Svechin would not have understood the idea of the strategic corporal. He was interested mainly in having commanders of various levels (fronts, armies, and corps) understand strategy. But the idea of keeping strategy in the vernacular and not restricting it to a small group at the top is sound.

Marcella and Fought argue that strategy should be taught at all levels of the military: company-grade schools, intermediate service colleges, and the war colleges.[25] There are three reasons why this is a useful suggestion. First, some officers will be in senior leadership positions and will contribute to the formulation of strategy. What no one can know in advance, however, is who will end up with this responsibility. As General John Galvin once asked, "[f]or each accomplished strategist we produce, how many must begin the long period of winnowing and development?"[26] The answer is that most officers who attend the military schools of PME should study strategy at a level appropriate for their rank. The curriculum will, of course, become more sophisticated and intense at the senior service schools. The second reason for studying strategy at all levels is that those who do not find themselves with responsibility for devising strategy may be involved in implementing it, and these officers also need to be able to think strategically. Finally, as Marcella and Fought note, younger officers are making strategic calculations within their domain, although we call that domain tactical or operational.

The war colleges are the institutions that have the clearest responsibility for preparing officers for strategic roles. In my discussion here I will focus on the U.S. Army War College (USAWC), although most of my comments apply to the other senior service colleges as well.

Models of Strategy.

In his contribution to this volume, Harry Yarger describes the pedagogical approach used at the USAWC to teach strategy.[27] He notes the longstanding distinction between the art and science of strategy. The art is ". . . the ability to see the strategic dots and connect them in a meaningful manner. . . ."[28] The science is the body of knowledge—history, international relations, diplomacy, economics, ethics, psychology, etc.—that informs the strategic art.[29] Formal education can provide a foundation for strategic intuition, but it cannot teach it directly. Instead, intuition—the art of strategy— is the product of individual study and reflection, the integration of prior learning, and, perhaps, a natural inclination to think in this manner. Nonetheless, professional military education has an obligation to provide the environment and the intellectual stimulation for serious study and continuing education beyond the formal school year. We are, of course, far removed from a society in which a lone genius can sketch out a winning strategy for organizations as complicated as the military services of the United States. Strategic intuition is not enough. To have an impact a strategist must also know how to inspire subordinates and superiors with carefully crafted statements, reports, testimonies, orders, and memos, give persuasive PowerPoint briefings, and operate in environments that are both bureaucratic and political. Modern bureaucratic states have many interests, levels, and institutions, and the only way to coordinate action is to have processes and organizational routines. This aspect of strategy *can* be taught. As Yarger writes, "[a] science of strategy suggests that we can study strategy formulation, theorize about it, and improve performance by better understanding the processes involved."[30]

Toward this end, for two decades the USAWC has taught strategy by using a model developed by Arthur F. Lykke, Jr. The key components of the model include:

- Ends (objectives),
- Ways (strategic concepts/courses of action),
- Means (resources),
- Risk (the gap between what is to be achieved and the concepts and resources available to achieve the objective). The strategist seeks to minimize this risk through his development of the strategy—the balance of ends, ways, and means.[31]

Such a model can help students identify important relationships between critical components of strategy. It provides a common vocabulary rooted in the rational actor model of human behavior, a model that reflects the problem solving nature of thought.

Without qualification and context, however, such a model can lead to an oversimplified view of reality. The process of formulating strategy is too complex and idiosyncratic to be captured adequately in any abstract model. It is prone to the vicissitudes of human imperfections, group think dynamics, and the opponent's reaction. Moreover, it suggests that strategy is risk free if resources balance the ends. In the real world, there is no risk free strategy. Students are, of course, intelligent enough to realize that there is a difference between models of strategy making and the real world. But once we start thinking in terms of categories, our thought processes can become hostage to mental maps in ways that we may not always realize.

An ends-means model could imply that strategy is actually devised in a rational sequence of ends, ways, and means. Although he was writing about strategic programs rather than strategy, Samuel Huntington pointed to the complicated world of policymaking when he noted that such programs do *not* result from policymakers ". . . rationally determining the actions necessary to achieve desired goals."[32] Instead, both strategic programs and strategy itself are affected by the bargaining and pulling and hauling that have been described in the literature on bureaucratic politics and organizational process.[33]

The best way to avoid the incorrect impression that strategy formulation is always rational is to embed an ends-means model in a larger context. A good way of doing this is to use the concept of strategic culture. One goal of the USAWC, for example, is for its students to understand the cultural aspects of strategy and policy. The Strategic Thinking core course includes a consideration of culture, and an analytical cultural framework for strategy and policy is used in other courses as well.[34] These components of the curriculum should be examined carefully to ensure that they provide an adequate introduction to strategic culture.

The fundamental question about any pedagogical model of strategy is whether it accurately captures the dynamics of conflict in the current environment. Is the Lykke ends-means model timeless? Does it apply to all periods of strategic history? Or was it more relevant to World War II and the Cold War than it is today? A strong argument can be made that the Lykke model was more relevant to the age of industrial warfare when America had unmatched resource and logistical capabilities. In the future, material superiority alone will not win wars. American strategists will need to use the full spectrum of assets, including both hard and soft power.

An ends-means model can lead to an overly mechanistic view of strategy, thereby reducing the vital human element. It is important for students to understand the dynamic, interactive, and inherently psychological aspects of war. The French general, André Beaufre, offered an influential way of thinking about this. He defined strategy as "the art of the dialectic of two opposing wills using force to resolve their dispute."[35] He noted that there will be available a wide variety of means ranging from physical force to propaganda to economic tools. The art of strategy, he wrote: "... consists in choosing the most suitable means from those available and so orchestrating their results that they combine to produce a psychological pressure sufficient to achieve the moral effect required."[36] Any model used to teach strategy should incorporate or be supplemented by explicit attention to these interactive and psychological dimensions.

If models of strategy formulation are used to teach basic concepts, essential relationships, and a common strategic vocabulary, they must be used carefully in a way that minimizes the chance that they will be reified. One way to emphasize the dynamic nature of strategy is to examine carefully chosen cases in strategic history.

Using Case Studies.

Case studies enable us to draw attention to some of the most important strategic situations. One issue that is central to the choice of cases is deciding what span of time is relevant. Can we still learn applicable lessons from the ancient world or the American Civil War, World War I, or World War II? Colin Gray argues that we can. "It really does not matter whether strategy is 'done' by 'foot, horse, and guns', or by cruise missiles, spacecraft, and cyber-assault." To Colin Gray, ". . . nothing essential changes because there is a unity to all strategic historical experience."[37] Although I am not certain about the unity of all strategic historical experience, I do believe that those who formulate strategy in the 21st century can learn from case studies from the 20th century and probably from previous times as well.

The reason we should not overemphasize contemporary cases is that we cannot know what types of conflict lie around the corner. To cite the most recent example, until the unexpected rise of insurgency in Iraq, the U.S. Army had paid little attention to counterinsurgency strategy in recent years. The way to avoid that situation is to have students in professional military education study historical cases far removed from preconceptions about future conflicts. Bernard Brodie once described the importance of reading widely in the following way:

> [m]eaningful parallels [are] usually not to be found in the leader's own experience, though he may have found them in a creative reading of history — the kind of reading that enables one without effort and perhaps only half consciously, or even unconsciously, to recall some past instance that bears in some significant way on a present problem.[38]

We need military officers who have read broadly about the past so that they will be able to apply their knowledge to unexpected conflicts with unanticipated characteristics.

Strategy is multidisciplinary and multidimensional. It is difficult to teach anyone how to formulate strategy, but it is surely impossible if the student is unaware of the challenges that previous strategists have faced. Even (or especially) for military officers with operational experience, the necessary starting point is studying the past.

Used correctly, case studies can be more than historical exercises. They can promote active learning to enhance a student's understanding of strategy formulation and implementation.[39] In selecting cases, it does not matter whether the strategy worked or not. A student can learn as much from examining the Vietnam War as from World War II.

American strategic planning in the 1930s offers an illustrative case of planning in peacetime. Although the United States was far from prepared for World War II in operational terms, the RAINBOW plans provided strategic direction. The final plan developed prior to the Japanese attack on Pearl Harbor, War Plan RAINBOW 5, ". . . set down the basic strategy of a global war before this country was involved in it"[40] The development of the RAINBOW plans is a useful case study of strategic planning during a time when much of the American public expected to avoid war.

A case that has been widely used as an introduction to *grand* strategy is the formulation in 1950 of NSC-68. The authors urged "a rapid build-up of political, economic, and military strength in the Free World."[41] They recommended a comprehensive and decisive program that would involve increased military spending, more military and economic assistance, dealing with the balance of payments issue, improved intelligence, a reduction in nondefense federal spending, increased taxes, development of internal security and civil defense programs, and increased use of psychological warfare and covert operations.[42]

Although NSC-68 did not provide an integrated plan to address each of these issues, its identification of a wide range of tools is consistent with Edward Mead Earle's concept of grand strategy. Paul Nitze, the chief architect of NSC-68, wrote that it ". . . addressed what I have considered throughout my career to be the fundamental question of national security: How do we get from where we are to where we want to be without being struck by disaster along the way?"[43] That is a reasonably good short-hand description of the purpose of strategy. The NSC-68 case study is one of the best academic experiences at the USAWC.

For a case study of unsuccessful strategic planning, the Vietnam War is useful. Although there are many elements of Vietnam strategy that can be studied, an examination of the failure of strategic assessment is one of the most helpful to strategists of the future. John Lewis Gaddis noted "a persistent inability to monitor performance." One reason for this was excessive reliance on "easily manipulated statistical indices."[44] Scott Sigmund Gartner studied American ground warfare strategy from 1966 through the Tet Offensive and highlighted the problems caused for the war effort by the dominant indicators of success used by the military (e.g., enemy weapons captured and enemy killed.[45] The work of Gaddis and Gartner illustrates the importance of choosing relevant indicators of strategic success and making every effort to ensure that the top leadership pays attention to those indicators. This case study demonstrates the need for a functioning feedback loop in any strategic plan.

The American reformulation of strategy during the Iraq War provides a useful case study of strategic adaptation. A war conceived as a quick invasion became instead a prolonged war of counterinsurgency.

Steven Metz described the initial American strategy in the following terms: "However laudable the overarching American objectives in Iraq, the United States was strategically and conceptually unprepared to realize them. We used flawed strategic assumptions, did not plan adequately, and had a doctrinal void. . . . American strategy was characterized by a pervasive means/end mismatch."[46] From 2005 to 2007, the U.S. military changed direction. Stability operations were made a core U.S. military mission. The 2006 *Quadrennial Defense Review* and *National Security Strategy* noted the irregular nature of the evolving war in Iraq. Counterinsurgency was restored to the curriculum of the Army's major schools. And the first new counterinsurgency Field Manual in 20 years was written.[47] A careful examination of how American political and military leaders formulated and implemented these and related changes provides a case study of strategic innovation in the midst of war.

One challenge in teaching about strategy is designing a curriculum that uses case studies to engender in students a strategic way of thinking but that resists simplistic analogies. It is important to make it explicit to students that there can be no mechanistic or formulaic lessons of history. Discrimination must be used in distinguishing historical analogies that are helpful from those that lead us astray.[48] As Richard Betts once reminded us, "[s]ensible strategy is not impossible, but it is usually difficult and risky, and what works in one case may not in another that seems similar."[49]

Comparing Grand Strategies.

A popular approach to grand strategy involves comparing several ideal types of possible strategies. Barry R. Posen and Andrew L. Ross published a compelling example of this approach in the 1990s in their article "Competing Visions for U.S. Grand Strategy."[50] They described and assessed four strategies: neo-isolationism, selective engagement, cooperative security, and primacy. In the summer of 2008, the Center for a New American Security published an updated survey of similar strategies in *Finding Our Way: Debating American Grand Strategy*.[51]

This is a useful way to introduce students to rival theories of grand strategy. For teaching strategy in professional military education, however, this approach may be less helpful. The main problem is that the argument for each strategy is too coherent and too logical. The cases are, as you might expect from an ideal type, too ideal.

During the Cold War, when there was one major challenge to American power, the United States used a containment strategy, although even then that policy did not inform every situation. For example, Soviet influence was not kept out of Cuba. In today's world, with no single overriding threat, it is difficult to imagine a President being comfortable with a single integrated strategy to address the diverse problems that we face. It is much more likely that a President will want to find a solution to each problem without the burden of consistency that would be imposed by explicit adoption of any of the grand strategies mentioned above.

Although models of grand strategy are useful in highlighting the assumptions of various options, they do not take into account the limits of policy coherence in Washington. Indeed, the search for a master plan at the level of grand strategy may be illusory. As Aaron L. Friedberg wrote after working as director of policy planning in the Office of the Vice President, "[t]he true aim of national strategic planning is heuristic; it is an aid to the collective thinking of the highest echelons of the government, rather than a mechanism for the production of operational plans."[52]

CONCLUSION

In educating officers to think strategically in their careers, the war colleges can acquaint them with history, provide frameworks for thinking about strategy, embed strategy formulation in the broader context of strategic culture, and provide opportunities to develop strategies via exercises and simulations. Not every student will master the science of strategy, and for reasons discussed earlier, no formal curriculum can consistently produce students who will excel at the *art* of strategy. The realistic goal, as Harry Yarger phrased it, is to have faculty and students ". . . create an environment in which all can learn at differing levels the science and art of strategy. Faculty members can build on what has gone before and lay a foundation of knowledge and habits of thought for continued learning and practice. . . ."[53]

A nation in America's position in the international system has to be prepared for contingencies both large and small. As President Barack Obama put it at the Naval Academy Commencement,

. . . we do not have the luxury of deciding which challenges to prepare for and which to ignore. We must overcome the full spectrum of threats — the conventional and the unconventional; the nation-state and the terrorist network; the spread of deadly technologies and the spread of hateful ideologies; 18th century-style piracy and 21st century cyber threats.[54]

It will not be easy to design strategies for this environment. The reason it is important to continually assess the teaching of strategy at the senior service colleges is that their students will be involved in this vital work. The graduates of the war colleges need to be able to think strategically so that they can assist the elected leaders and civilian appointees of the government in meeting the many challenges facing the United States.

ENDNOTES - CHAPTER 3

1. Terry L. Deibel, *Foreign Affairs Strategy: Logic for American Statecraft*, New York: Cambridge University Press, 2007, p. 10.

2. Colin S. Gray, *Modern Strategy*, New York: Oxford University Press, 1999, p. 17.

3. Gabriel Marcella and Stephen O. Fought, "Teaching Strategy in the 21st Century," *Joint Force Quarterly*, Issue No. 52, 1st quarter 2009, p. 57.

4. Deibel, p. 9.

5. Edward Mead Earle, ed., *Makers of Modern Strategy: Military Thought from Machiavelli to Hitler*, Princeton, NJ: Princeton University Press, 1943, p. viii.

6. André Beaufre, *An Introduction to Strategy*, New York: Frederick A. Praeger, 1965, p. 31.

7. Justin Ewers, "Why Don't More Colleges Teach Military History?" *U.S. News and World Report*, April 3, 2008, available from *www.usnews.com/articles/news/2008/04/03/why-dont-colleges-teach-military-history.html*. Also see Patricia Cohen, "Great Ceasar's Ghost! Are Traditional History Courses Vanishing?" *New York Times*, June 11, 2009 available from *www.nytimes.com/2009/06/11/books/11hist.html*.

8. The 15 are, from number 1 to number 15, Amherst, Williams, Swarthmore, Wellesley, Middlebury, Bowdoin, Pomona, Carleton, Davidson, Haverford, Claremont McKenna, Vassar, Wesleyan, Grinnell, and Harvey Mudd.

9. There are, of course, a few excellent courses on strategy in American higher education. The most extraordinary one of which I am aware is the graduate level course offered in the Brady-Johnson Program in Grand Strategy at Yale University. Taught by John Lewis Gaddis, Paul Kennedy, and Charles Hill, the course extends over the spring, summer, and the fall semesters. For an excellent description of the course, see John Lewis Gaddis, "What is Grand Strategy?" keynote address for the conference on American Grand Strategy After War, February 27-28, 2009, Duke University, available from *www.pubpol.duke.edu/centers/tiss/DebatingGrandStrategyDetails.php*.

10. John Lewis Gaddis, *Strategies of Containment: A Critical Appraisal of American National Security Policy During the Cold War*, Revised and Expanded Ed., New York: Oxford University Press, 2005 (originally published in 1982); Peter Paret, ed., *Makers of Modern Strategy: From Machiavelli to the Nuclear Age*, Princeton, NJ: Princeton University Press, 1986; Williamson Murray, MacGregor Knox, and Alvin Bernstein, eds., *The Making of Strategy: Rulers, States, and War*, New York: Cambridge University Press, 1994. There are, of course, many books that deal with various dimensions of war without a specific focus on strategy. The book reviews in *The Journal of Military History* provide one of the best ways to stay abreast of the emerging literature.

11. On the early strategists and the RAND Corporation, see Fred Kaplan, *The Wizards of Armageddon,* New York: Simon and Schuster, 1983; and Marc Trachtenberg, *History and Strategy,* Princeton, NJ: Princeton University Press, 1991, pp. 3-46. For the most complete survey of writings on nuclear strategy, see Lawrence Freedman, *The Evolution of Nuclear Strategy,* Third Ed., New York: Palgrave Macmillan, 2003. For assessments of strategic studies and security studies, see Colin S. Gray, *Strategic Studies: A Critical Assessment,* Westport, CT: Greenwood Press, 1982; Stephen M. Walt, "The Renaissance of Security Studies," *International Studies Quarterly,* Vol. 35, No. 2, June 1991, pp. 211-239; Edward Kolodziej, "Renaissance in Security Studies? Caveat Lector!," *International Studies Quarterly,* Vol. 36, No. 4, December 1992, pp. 421-438; and Richard K. Betts, "Should Strategic Studies Survive?" *World Politics,* Vol. 50, October 1997, pp. 7-33. For the most comprehensive recent overview of the field, see Edward Kolodziej, *Security and International Relations,* New York: Cambridge University Press, 2005.

12. Dan Reiter and Curtis Meek, "Determinants of Military Strategy, 1903-1994: A Quantitative Empirical Test," *International Studies Quarterly,* Vol. 43, 1999, pp. 365-368.

13. Barry R. Posen, *The Sources of Military Doctrine: France, Britain, and Germany Between the World Wars,* Ithaca, NY: Cornell University Press, 1984; Jack Snyder, *The Ideology of the Offensive: Military Decision Making and the Disasters of 1914,* Ithaca, NY: Cornell University Press, 1984; John J. Mearsheimer, *Conventional Deterrence,* Ithaca, NY: Cornell University Press, 1985.

14. Stephen M. Walt, "Rigor or Rigor Mortis? Rational Choice and Security Studies," *International Security,* Vol. 23, No. 4, Spring 1999, p. 8. This article led to vigorous exchanges in subsequent issues. See, for example, Lisa L. Martin, "The Contributions of Rational Choice: A Defense of Pluralism," *International Security,* Vol. 24, No. 2, Fall 1999, pp. 74-83.

15. Alexander Wendt, *Social Theory of International Politics,* New York: Cambridge University Press, 1999, p. 1.

16. *Ibid.*, p. 1. For a broadly similar approach with a specific focus on national security, see Peter J. Katzenstein, ed., *The Culture of National Security: Norms and Identity in World Politics*, New York: Columbia University Press, 1996.

17. Elizabeth Kier, *Imagining War: French and British Military Doctrine Between the Wars*, Princeton, NJ: Princeton University Press, 1997.

18. Steven E. Miller, "*International Security* at Twenty-five," *International Security*, Vol. 26, No. 1, Summer 2001, p. 13. Another steady source of policy relevant research is the Adelphi Papers series of the International Institute for Strategic Studies (IISS). These have been published since 1961 and thus provide a window into many of the major topics in the field over the past five decades. For some of the more notable examples, see Patrick M. Cronin, ed., *The Evolution of Strategic Thought: Adelphi Paper Classics*, New York: Routledge, 2008.

19. Maurice Matloff, "Allied Strategy in Europe, 1939-1945," in Peter Paret, ed., *Makers of Modern Strategy*, pp. 677-702.

20. Peter D. Feaver and Richard H. Kohn, eds., *Soldiers and Civilians: The Civil-Military Gap and American National Security*, Cambridge, MA: MIT Press, 2001. For a discussion of the consequences of ineffective civil-military relations for strategic assessment, see Risa A. Brooks, *Shaping Strategy: The Civil-Military Politics of Strategic Assessment*, Princeton, NJ: Princeton University Press, 2008.

21. Many of the most successful officers do have a civilian educational experience. In the USAWC class of 2009, 238 of the 336 students had graduate degrees when they entered the program, and most of these degrees were earned at civilian universities. See Major General Robert M. Williams, "Statement before the Oversight and Investigations Subcommittee, House Armed Services Committee," U.S. House of Representatives, 1st Sess., 111th Cong., June 4, 2009, p. 16. For an interesting debate about the desirability of officers attending civilian graduate schools, see the exchange titled "To Ph.D. or Not to Ph.D . . ." in *The American Interest*, July/August 2007. Drawing on his own experience, David H. Petraeus presents a strong case in favor of civilian education in

"Beyond the Cloister," pp. 16-20. In "Learning to Lose," pp. 20-28, Ralph Peters opposes civilian graduate school on the grounds that such an education will turn officers into indecisive Hamlets.

22. Eliot A. Cohen makes a similar point in "Strategy: Causes, Conduct, and Termination of War," in Richard Shultz, Roy Godson, and Ted Greenwood, eds., *Security Studies for the 1990s,* Washington, DC: Brassey's [US], 1993, p. 79.

23. Peter Paret, "Introduction," in Peter Paret, ed., *Makers of Modern Strategy,* p. 8.

24. Aleksandr A. Svechin, *Strategy,* Minneapolis, MN: East View Publications, 1992, p. 74, quoted in Andrei A. Kokoshin, *Soviet Strategic Thought, 1917-91,* Cambridge, MA: MIT Press, 1998, p. 25. Svechin's book was first published in 1926.

25. Marcella and Fought, "Teaching Strategy in the 21st Century," pp. 57-58.

26. John R. Galvin, "What's the Matter with Being a Strategist?," *Parameters,* Vol. XIX, No. 1, March 1989, reprinted in Arthur R. Lykke, Jr., ed., *Military Strategy: Theory and Application,* Carlisle, PA: U.S. Army War College, 1993, p. 13.

27. Harry R. Yarger, "How Do Students Learn Strategy?" in this book, pp. 179-202. Major General Robert Williams, Commandant of the U.S. Army War College, has also described the curriculum recently in the testimony before the House Armed Services Committee cited earlier.

28. Harry R. Yarger, *Strategy and the National Security Professional: Strategic Thinking and Strategy Formulation in the 21st Century,* Westport, CT: Praeger Security International, 2008, p. 6.

29. Although discussing the science of strategy is common, there are components of strategy that are far from scientific. History, diplomacy, and ethics, for example, do not use the scientific method. For a skeptical view of whether strategy can be viewed as a science, see John Gooch, "History and the Nature of Strategy," in Williamson Murray and Richard Hart Sinnreich, eds., *The Past as Prologue: The Importance of History to the Military*

Profession, New York: Cambridge University Press, 2006, pp. 138-139.

30. Yarger, "How Do Students Learn Strategy?," p. 179.

31. Harry R. Yarger, "Toward A Theory of Strategy: Art Lykke and the U.S. Army War College Strategy Model," in J. Boone Bartholomees, Jr., ed., *U.S. Army War College Guide to National Security Issues. Vol. I: Theory of War and Strategy*, Carlisle, PA: Strategic Studies Institute, U.S. Army War College, 2008, p. 47. A similar approach to formulating strategy was used at the National War College in the 1980s. William J. Stewart developed a model whose key concepts were objectives, capabilities, assumptions, and costs. William J. Stewart, "Strategy: A Proposed Model for Its Formulation," in George Edward Thibault, ed., *The Art and Practice of Military Strategy*, Washington, DC: National Defense University, 1984, pp. 7-14.

32. Samuel P. Huntington, "Strategic Planning and the Political Process," in Davis B. Bobrow, ed., *Components of Defense Policy*, Chicago, IL: Rand McNally, 1965, p. 84.

33. *Ibid.*

34. General Robert Williams, House Armed Services Committee, pp. 11-12. Although strategic culture is a helpful pedagogical device, it must be used carefully. In scholarly research the concept is controversial. Colin Gray was an early exponent of the approach. See his article "Comparative Strategic Culture," *Parameters*, Winter 1984, pp. 26-33. For an interesting exchange reflecting the debate about the concept, see Alastair Iain Johnston's critical view in "Thinking About Strategic Culture," *International Security*, Vol. 19, No. 4, Spring 1995, pp. 32-64; and Colin Gray's response in "Strategic Culture as Context: The First Generation of Theory Strikes Back," *Review of International Studies*, Vol. 25, 1999, pp. 49-69.

35. Beaufre, *Introduction to Strategy*, p. 22.

36. *Ibid.*, p. 24.

37. Gray, *Modern Strategy*, p. 358.

38. Bernard Brodie, *War and Politics*, New York: Macmillan Publishing Co., 1973, p. 435.

39. Vicki L. Golich, Mark Boyer, Patrice Franko, and Steve Lamy, *The ABCs of Case Teaching*, Washington, DC: Georgetown University Institute for the Study of Diplomacy, 2000. Case studies are also useful for research purposes. See Alexander L. George and Andrew Bennett, *Case Studies and Theory Development in the Social Sciences*, Cambridge, MA: MIT Press, 2005.

40. James J. Schneider, "War Plan RAINBOW 5," *Defense Analysis*, Vol. 10, No. 3, December 1994, pp. 285-304. For a book length discussion, see Edward S. Miller, *War Plan Orange: The U.S. Strategy to Defeat Japan, 1897-1945*, Annapolis, MD: Naval Institute Press, 1991.

41. Ernest R. May, ed., *American Cold War Strategy: Interpreting NSC 68*, Boston, MA: Bedford Books of St. Martin's Press, 1993, p. 71. Also see S. Nelson Drew, ed., *NSC-68: Forging the Strategy of Containment*, Washington, DC: National Defense University, 1994.

42. May, *American Cold War Strategy*, p. 74.

43. Paul H. Nitze, *From Hiroshima to Glasnost: At the Center of Decision*, New York: Grove Weidenfeld, 1989, p. 95. Although NSC-68 is important as an exercise in grand strategy, its predecessors deserve attention as well. The 1948 document NSC 20/1, for example, clearly identified not only war aims in the event of war with Russia *but also* "those things which we could not hope to achieve." Clarity on what lies beyond the scope of military action should be a component of any strategic plan. See "U.S. Objectives with Respect to Russia," NSC 20/1, August 18, 1948, in Lawrence Freedman, ed., *War*, New York: Oxford University Press, 1994, pp. 291-297.

44. Gaddis, *Strategies of Containment*, p. 253.

45. Scott Sigmund Gartner, *Strategic Assessment in War*, New Haven, CT: Yale University Press, 1997, pp. 117-146.

46. Steven Metz, *Learning from Iraq: Counterinsurgency in American Strategy*, Carlisle, PA: Strategic Studies Institute, U.S. Army War College, 2007, p. 85. Also see his expanded discussion of this in *Iraq and the Evolution of American Strategy*, Washington: Potomac Books, 2008.

47. Metz, *Learning from Iraq*, pp. 61-65.

48. For discussions of applying — or misapplying — the lessons of history, see Richard E. Neustadt and Ernest R. May, *Thinking in Time: The Uses of History for Decision Makers*, New York: Free Press, 1986; and Jeffrey Record, *Making War, Thinking History: Munich, Vietnam, and Presidential Uses of Force from Korea to Kosovo*, Annapolis, MD: Naval Institute Press, 2002.

49. Richard K. Betts, "Is Strategy an Illusion?" *International Security*, Vol. 25, No. 2, Fall 2000, pp. 5-50. The quotation is on page 48.

50. Barry R. Posen and Andrew L. Ross, "Competing Visions for U.S. Grand Strategy," *International Security*, Vol. 21, No. 3, Winter 1996/97, pp. 5-53.

51. Michele A. Flournoy and Shawn Brimley, *Finding Our Way: Debating American Grand Strategy*, Washington, DC: Center for a New American Security, 2008. For book length discussions of grand strategy, see Robert J. Art, *A Grand Strategy for America*, Ithaca, NY: Cornell University Press, 2003; and Christopher Layne, *The Peace of Illusions: American Grand Strategy from 1940 to the Present*, Ithaca, NY: Cornell University Press, 2006. Art considers eight grand strategies and argues for selective engagement. Layne argues for offshore balancing.

52. Aaron L. Friedberg, "Strengthening U.S. Strategic Planning," *The Washington Quarterly*, Vol. 31, No. 1, Winter 2007-08, p. 48. Friedberg paints a bleak picture of the capacity of the U.S. Government to engage in serious strategic planning at the level of grand strategy. For an interesting and useful discussion of strategic planning with a focus on counterterrorism , see Lynn E. Davis and Melanie W. Sisson, *A Strategic Planning Approach: Defining Alternative Counterterrorism Strategies as an Illustration*, Santa Monica, CA: RAND Corporation, 2009.

53. Yarger, "How Do Students Learn Strategy?" p. 199. For some useful suggestions for reforming the war college curriculum, see the chapter by Marcella and Fought in this volume.

54. Remarks by the President at the U.S. Naval Academy Commencement, May 22, 2009, *available from www.whitehouse.gov/ the_press_office/Remarks-by-the-President-at-US-Naval-Academy-Commencement/*.

CHAPTER 4

TEACHING STRATEGY IN THE 21ST CENTURY

Gabriel Marcella
Stephen O. Fought

THE WAR COLLEGE ACADEMIC EXPERIENCE[1]

The war colleges of the United States—Army, Navy, Air Force, National, Industrial College of the Armed Forces, and Marine Corps—are unique national assets. Their missions are similar: to prepare military and civilian officers, American and international, for responsibilities of strategic leadership in joint, interagency, and multinational positions in Washington, DC, and in the field. The nation makes a heavy investment in professional military education at the war colleges. Each year they graduate nearly 1,500 mid-level officers to replenish the talent pool of the armed services. Graduates will assume duties in such locations as the National Security Council staff, State and Defense Departments, Homeland Security, the Joint Staff, the Services, combatant commands, embassies, multinational commands, and organizations such as the United Nations (UN). Such assignments demand wisdom in the art of making strategy and advising about the use of military power in support of American foreign policy. Accordingly, war colleges are living laboratories for studying how to use power for political purpose.

Because of the high level impact of the education and the high quality of students, there is nothing equal to the intellectual delights of mutual discovery via the adult active learning method, the Socratic give and take in seminars at the various colleges. The seminar, the preeminent form of instruction, is composed of 14-16 students and deliberately diverse in terms of military service, specialties, gender, race, and international officers. Students, military and civilian, are accomplished, demanding, talented, and interesting professionals who will ascend to the highest positions in the national security system.[2] For military students, attendance at a war college is the last required gate for promotion eligibility to general officer. Instructors are productive scholars who teach, write, and interact with the policy-strategy communities in Washington, with American society, and in the international community.

The civil-military quality of the teaching and research faculties are models of professional collaboration while the opportunities to contribute to our nation's defense by preparing the nation's future military and civilian leaders are compelling challenges. Indeed, it is common for professors to circulate between teaching assignments and the policy-strategy communities in Washington and in the field. Similarly, senior diplomats, as well as civil servants from the defense and intelligence communities, normally rotate to teaching positions at the war colleges.

War colleges have become centers of academic excellence, enriched by numerous initiatives, such as the inclusion of foreign officers, increased participation by civilian students from such departments as State, Defense, and other agencies, the growth of a professional faculty, and the productivity of scholars whose publications reach the national and international marketplace of strategic studies and the highest levels of government in Washington and abroad. In the last 10 years, all the war colleges have developed a

Master's program in strategic studies, which is a requirement for students to graduate.[3] The extensive outreach to American society and the global community further enriches the academic programs and strengthens the democratic character of civil-military relations.

The colleges have similar curricula, though they differ in emphasis, and in methods. Obviously, land power is the emphasis at the U.S. Army War College, naval power at the Naval War College, air power and technology at the Air War College, and national security and the interagency (especially the Department of State) at the National War College. The young and small Marine Corps War College has a curriculum that tracks with the other schools. The curriculum is updated each year to adjust to changing emphases in military strategy, and to take into account new literature. In recent years there has been greater focus on joint operations, the whole of government, interagency approach to national security, as well as the new threats on the global scene. In doing so, the colleges balance the need to be timely while maintaining the conceptual foundation of strategic thought. Service culture and tradition powerfully influence how each school does things, including teaching. In the spirit of letting a thousand flowers bloom, each war college teaches differently, though the seminar form of instruction dominates. The Naval War College, for example, uses the case study method more than the other schools. In addition, the Naval War College, U.S. Army War College, Air War College, and Marine War College are truly residential colleges, allowing for a rich learning environment. In contrast, the National War College and the Industrial College of the Armed Forces are commuter schools, with limited time on campus for interaction with faculty and student colleagues.

War colleges (also known as senior service colleges) have matured as educational and research institutions. Nonetheless, they need to do a much better job at the core mission of teaching strategy. There is a growing concern that they teach about strategy, rather than teaching about how to develop strategy. This concern became paramount because of the failure of the military to plan an effective strategy for the war in Iraq. But the problem goes beyond the current headlines. The root of the problem is the a-strategic nature of American society and the reliance on resource superiority to win wars in the past.

The intense 10-month curriculum, which leads to the Master's degree, includes a full plate of leadership and management, theories of war and strategy, the military budget, the national security decisionmaking process, strategic and operational planning, some international relations and foreign policy, civil-military relations, military history, the instruments of power, campaign planning, current and future threats, and the study of the various regions of the world. Each war college runs a distance education program of 2-years length, whereby officers conduct their studies while working full time. This is a highly demanding program.

The curriculum acquaints students with various national strategy documents and offers a menu of electives that expand intellectual horizons, including a dose of understanding foreign societies and cultures so that officers can appreciate and be able to work more effectively within the historical, political, religious, and economic contexts of both allies and adversaries. Such knowledge is critical in the 21st century because officers will have to deal with the globalization of the

Clausewitzian trinity: the people, the government, and the armed forces of the multiple states that will be U.S. partners or opponents in a variety of enterprises that will require the use of the military instrument in combination with all tools in the quiver of national power.

STRATEGY AND FUTURE ASSIGNMENTS

War college graduates will be expected to make strategy regardless of their military specialties. Graduates entering joint assignments, combatant commands, or multinational forces/joint task forces will be involved in making or shaping strategy via their interaction with planners. These officers will provide advice on behalf of their organizations during the development of strategy. At the grand strategic level, they will be desk officers, military advisors, or staffers working at places like the National Security Council staff, State, Defense, Homeland Security, the Joint Staff, combatant commands, Service staffs, the UN, and coalition organizations. Graduates will advise senior leaders and will be expected to write extensively and well. Additionally, they will be expected to analyze strategic level guidance from a variety of sources to determine executive and legislative branch intent. For example, officers assigned to the Office of Secretary of Defense, the Joint Staff, or the interagency in support of civilian leaders may work as part of a team writing the National Security Strategy, National Defense Strategy, National Military Strategy, National Counterterrorism Strategy, and other documents. These assignments call for mastery of the nuances of policy and strategy.

War college graduates will also work at the strategic-operational level in combatant commands, multinational commands, or joint task forces. Examples include Multinational Force-Iraq, Multinational Corps-Iraq, or Joint Task Force-Horn of Africa. They will be working with joint, interagency, and coalition partners. At the combatant commands, they will often have to divine strategic level guidance from legislative and executive branch documentation to provide concise consolidated guidance to planners. The powers of divination require a comprehensive understanding of how the U.S. Government works, in addition to diplomacy and articulation skills in writing and speaking. Graduates will give life to strategy by extracting that guidance, clearly communicating it to leaders and peers, and "drawing the box" within which they will plan.

Moreover, officers need to understand that non-executive branch guidance is fundamental to their work. At times this guidance may not exist in any detail and must therefore be once again "divined" through research and consultation with subject matter experts. At the joint task force or multinational force level, graduates are expected to discuss strategic level guidance in detail. As an example, the officer who provides the strategic scene-setting for American support to Colombia must seek guidance not only in national level policy and strategy documents, but also in legislation that goes back a decade or more. He or she should also be intimately familiar with the various players in Washington who have a role in Colombian affairs, in counternarcotics, counterterrorism, intelligence, governance, post-conflict reconstruction, and the rule of law. Another example is that officers in the American Embassy Baghdad must be able to assist State Department and other interagency planners in the Joint Strategic Plans and Assessments Directorate with finding, analyzing, and integrating strategic level guidance for the proper use of military power.

THE CHALLENGE OF TEACHING STRATEGY

The war colleges teach by utilizing a variety of pedagogical techniques—lectures, seminars, group work, simulations, and writing research papers. The objective is to expose students to a variety of information and organizing tools so that they become better strategists. Some students may put this learning together and become better at synthesis, a critical element of the art of strategy. But success is not certain. Moreover, recent performance by American civilian and military leaders in the difficult war in Iraq and the longer war in Afghanistan leads to the conclusion that the schoolhouse must do a better job of teaching. There are many hurdles to overcome, not the least of which is the pressure of time. The tyranny of the military personnel assignment system forces the schoolhouse to squeeze in a lot of course work within the 10-month calendar so that officers can assume command and staff duties, because demand for staff officers and commanders is very high. High demand has been especially true since the 1990s, given a smaller military and more missions to perform. Within this period of time academic work equivalent to the master's degree must be delivered, which entails 30 credits of inquiry and writing. A master's program, with or without thesis, is normally of 2 years length. Getting it done in 10 months engenders a very intense pace for professionals in their 40s who have children and elderly parents and are planning their next assignments in a highly structured career path that values command and leadership. Lamentably, the equally intense operational tempo of the U.S. military since the end of the Cold War has also limited the time for self-generated study, reading, and reflection. It is in this busy context that strategy must be learned.

Triage with the curriculum is inevitable and welcome. But this is not easy to do, in part because of requirements levied by outside organizations, such as the Joint Staff, the Department of Defense (DoD), and even the Congress. For example, Representative Ike Skelton of the House Armed Services Committee has a deep interest in the quality of professional military education. In this remarkable example of co-responsible democratic civil-military relations, he and his staffers regularly visit the war colleges, require briefings on the academic programs, and follow up with hearings before the subcommittee entrusted with oversight on military education—the Oversight and Investigations Subcommittee.[4] Moreover, the creation of the master's program in 2000 benefited the war colleges by instilling a level of academic rigor that must be adhered to in order to maintain accreditation. Given the vicissitudes described above, some have proposed a 2-year war college program, but this would require serious adjustments in the personnel system.[5]

DEFINING STRATEGY AND POLICY

The discussion of teaching strategy effectively must begin with definitions. Richard K. Betts defines strategy:

> as a plan for using military means to achieve political ends . . . a value added to power. Strategy is the essential ingredient for making war either politically or morally tenable. It is the link between military

means and political ends, the scheme for how to make one produce the other. Without strategy, there is no rationale for how force will achieve purposes worth the price in blood and treasure. Without strategy, power is a loose cannon and war is mindless. Mindless killing can only be criminal. Politicians and soldiers may debate which strategic choice is best, but only pacifists can doubt that strategy is necessary.[6]

He states that strategy is a multilayered chain of relationships between policy and power, spanning grand strategy, foreign policy, national military strategy, theater strategy, war, campaigns, and engagements.[7] André Beaufre (1902-75), the French military strategist, offered another insight. He stated that "strategy is the art of the dialectic of two opposing wills using force to resolve their dispute."[8] He argued that the strategic objective can be attained through two modes of strategic behavior: direct and indirect. In the first, the military instrument dominates, while in the indirect mode, the military instrument becomes a supporting secondary means. Both modes are total strategy, involving the integration of all the instruments of state power. Another critical insight by Beaufre is the interaction of power and will, which he called the moral factor. This allows us to factor such cognitive dimensions as leadership, psychology, intelligence, group dynamics, values, and the cognitive domain of the opponent, in the development and conduct of strategy. These valuable insights should inform the teaching of strategy.

In tune with the multi-instrument nature of power and the needs of the twenty first century, we propose this definition: *Strategy is the art of applying power to achieve objectives, within the limits imposed by Policy.* Note the emphasis on art, rather than science, a matter that will be treated later in this chapter.

The definition of Strategy begs for a definition of Policy, since the latter controls the former: "Policy is the guiding principle to justify, and bound, the application of resources and effort over time to achieve objectives that promote the national interests of the nation, such as defense, economic prosperity, international order, and protection of human rights."[9] The planning and implementation of strategy, especially in wartime, requires, to paraphrase Clausewitz, a permanent conversation with policy. The subordination of policy to strategy, the domination of ends by means, is a risk that must be averted. It is a prescription for disaster of the kind seen in World War I and again in Vietnam. Disaster can take another form, where operations are conducted in the absence of strategy: Korea and the second Iraq war. If policy is the *what*, strategy is the *how*. It is imperative that the statesman/strategist sort out the two, for history is replete with their confusion.

There is another critical aspect to policy: it limits strategy. Strategy is developed and exists at every level, the tactical, operational, and the very strategic. Strategy is developed with the purpose of connecting objectives to military means. Limits to freedom of action exist at every level and must be accounted for by those who develop strategy. These limitations are, collectively, called policy. When one develops strategy, one develops limits (hence policy) on other levels — on levels below, certainly, on collateral levels, quite often, and on levels above, sometimes by accident, what Clausewitz might have called the fog of strategy.

STRATEGY AS AN INTEGRATOR

Strategy at all levels, from grand strategy to military strategy, is integrative. It pulls together various strands

to make the whole cloth. There are at least 13 ways in which strategy is an integrative. These are eminently teachable elements of the strategy curriculum.
Strategy:

1. Is designed to influence and control the political behavior of the adversary, perhaps to the point of exercising control in order to modify the power structure of a society and the regional geopolitical balance favorably.

2. Integrates political purpose with military means.

3. Integrates conceptualization, budgeting, and implementation.

4. Integrates joint, interagency, and coalition capabilities.

5. Integrates the instruments of national power through all the phases of war, from pre-conflict, to operations, conflict termination, post-conflict reconstruction and stabilization, and securing sustainable peace.

6. Integrates theater strategy with regional and global priorities.

7. Integrates realism with idealism by linking political purpose, military means, and moral authority (*jus ad bellum* and *jus in bello*).

8. Integrates multiple disciplines: foreign policy, politics, international relations, psychology, law, ethics, economics, and anthropology.

9. Integrates the wisdom of the past (e.g., Thucydides, Clausewitz, Machiavelli, Sun Tzu, Mahan, Liddell Hart) — the "science" of strategy with the "art" of making and applying strategy in the circumstances at hand.

10. Integrates Clausewitz's trinitarian components of passion (people), rationality (government), and military competence (armed forces).

11. Integrates the strategic with the operational and tactical levels, distinctions that are increasingly less meaningful in the 21st century.

12. Adjusts to the cultural norms and codes of conduct of foreign societies.

13. Integrates civilian and military capabilities in strategy and operations.

STRATEGY: SCIENCE VS. ART

There is the science and the art of Strategy. David Jablonsky, prolific writer and instructor for a generation of U.S. Army War College students, cautioned: ". . . students weaned on the structural certitude of the five-paragraph field order and the Commander's Estimate naturally find . . . structure comforting when dealing with the complexities of strategy.[10] He advised "In an ever more interdependent world in which variables for the strategist . . . have increased exponentially, strategists are no nearer to a 'Philosopher's Stone' than they ever were. Strategy remains the most difficult of all art."[11] The easy part is learning the wisdom of the great philosophers of war and strategy, such as Clausewitz, Sun Tzu, and Liddell Hart. This is the science (from the Latin *scientia*), the corpus of knowledge handed down by the masters. The challenge for the strategist is to apply such science to the art of making strategy in the crucible of modern conflict—precisely where schools of higher military education must make their contribution.

There is no recipe and no doctrine for making strategy or strategists. Pericles, Bismarck, Churchill, and their like possessed innate genius, seasoned by experience (to include failure) and self-study.[12] The American way of war, which historically relied on plentiful resources, technology, kinetics, and geographic cushioning, predisposes us against strategic creativity.[13] In fact, in the past our enormous advantage in resources masked flaws in strategy. Colin Gray remarks that Americans go straight from policy to operations, skipping strategy. Teaching strategy often meant training how to use resources for the various levels of the spectrum of conflict. Kinetics and resource superiority alone will not win future wars, while the geographic cushion will no longer save us from menace.

There is very little literature on this, unless one defines the training doctrine for operations and tactics as education. Training and education are polar opposites. The former is designed to eliminate the possibility of error through repeated practice, while the latter is designed to expand options for the decisionmaker/strategist. The available doctrine has to do with leadership, organization, logistics, intelligence, and operations, but offers nothing on the teaching and making of strategy.[14] The recent transformation, with its emphasis on the sinews of military power and technology, has further orphaned strategy. Most of the literature leaves us with a rich lode of theory and the use of military history to guide us in strategic thinking for the future. Thus, we must rely on proven pedagogical techniques, such as case studies that encompass both success and failure, mentoring, self-study, and writing.

DIALOGUE BETWEEN POLICY AND STRATEGY

We must teach students how to translate policy guidance into military strategy. To aid this process, Colin Gray echoes Clausewitz in advocating a permanent dialogue between policymaker and soldier, between policy and strategy.[15] There are some fundamental questions that should be inserted into the dialogue between policy and strategy:[16]

1. What national interests are engaged in the issue and how?

2. What are the threats?

3. What are the desired objectives and the level of commitment to those objectives by the government and American people? What are the U.S. options for instruments of national power? Are the national interests, objectives, and commitments articulated clearly and are they in balance? What are the costs of action vs. inaction?

4. What is the threshold, the spectrum of options for the use of military power? What is the military strategy? Are the objectives in the strategy mutually supportive and fully congruent with national interests? Is the strategy feasible and suitable, and are the costs acceptable?

5. What is the level of coordination-integration among the executive departments and agencies and the instruments they wield?

6. Are the American people and the Congress supportive? Are the resources adequate to the task? If not, what corrections in strategy are required?

7. What are the indicators that the strategy employed is achieving desired objectives? Should midcourse corrections be necessary and how will they be made?

8. What is the war termination and post-conflict reconstruction strategy?

9. Are allies and regional governments supportive and in what capacity are they contributing capabilities?

10. Are operations synchronized with strategy, and ultimately with policy? Which one dominates?

Though these questions may look pristine and clear, in the real world the policy guidance may be murky or nonexistent. There will be ambiguity, perhaps contradiction, and certainly competing priorities and competing values for the use of American military power. To reiterate, the power of "divination" will be useful for the strategist. He/she may have to divine policy guidance from insufficient information, incorrect assumptions, and perhaps insufficient resources.

THE PASSION OF STUDENTS: OUR GREATEST RESOURCE

War college students are, by virtue of being confident and successful commanders and managers of great resources, passionate about synthesis, of putting everything together to solve problems. Pedagogical theory holds that such adult professionals, forged in the crucible of pragmatic real world experience, need to see the application of learning. Moreover, learning is a social activity. Learning is also contextual and cumulative: we learn based on what we know, believe, and fear.[17] Failure and frustration are great teachers, success merely reinforces what we know and may set us up for failure. Finally, we understand organizing principles better as we use them. Co-author Steve Fought argues that war college students are impatient with theory. Therefore, they should confront problems to solve early on in the curriculum:

> They want the problem—now. So begin with a problem that stretches their capabilities, and let them flail. As flailing becomes failing, offer up theory to get them back on track. At some point, sometimes after they have hosed up the exercise completely, one of them will sheepishly ask: Has anybody ever

done this before? . . . Talented, experienced adults are aggressively impatient. They demand proof of relevance. The best method of proof is not to "show them" but to have them convince themselves. The roadmap is application-theory-history, offered in seminar environment, through real-world cases, accompanied by active student participation in both the learning and teaching processes.[18]

Fought's unstated assumption is that students are fairly well-grounded in the profession of arms. Hence they are eager to try their hands at larger problems, notably national security and military strategy. Therefore, we should not be captive to the order of march of theory-history-application for the curriculum. The order of presentation should take full advantage of the knowledge, experience, and passion for solving the nation's problems that students bring to the classroom.

GUIDELINES FOR CURRICULUM

Below are some initiatives in building a curriculum for the 21st century. They should be taken *in toto* as a comprehensive approach to strategic pedagogy.

1. Restructure the curriculum sequence to:

a. Introduce the students to the national security threats and challenges facing the nation. Require them to come up with creative solutions.

b. Once they have been presented with the threats and challenges and allowed a reasonable time to grapple with them, expose them to the theories, tools of analysis, and historical and contemporary case studies.

c. As students attempt to solve the national security and military challenges facing the nation, have them develop the military strategy, acquire the appropriate budget, build the force and the instruments of power, and exercise them in credible scenarios.

2. Develop an integrated strategy model as a pedagogical tool that can be applied to illustrate how all the instruments of national power are fused in the development and implementation of strategy at the various levels of peace and conflict. This should not be a mere flow chart of boxes and arrows, but rather fully developed writing on how strategy is made, in order to illustrate the nonlinear intellectual, cognitive human dimensions. We need to teach the diplomatic, informational, military, and economic (DIME) elements of power as integrated strategy, not as discrete elements simply tossed into the crucible when the military instrument is found wanting. The best minds should be tapped for this, including mentorship from creative strategic leaders, civilian as well as military. Mentors from the ranks of strategists should be inserted into war college seminars to enrich the learning process. They should challenge the assumptions of students.

3. Develop strategy components in the core curriculum, where students would be required to develop strategy for contemporary national security and military problems. Students should develop a national security strategy, followed by a military strategy that would have to be budgeted and then applied to the real world. The intellectual challenge of developing national security strategy engenders the skill of thinking holistically, a talent that can be applied to developing military strategy. If students simply analyze, as they currently do, published strategy documents (such as the National Security Strategy and the National Military Strategy) that are written by professionals experienced in statecraft, they are spared the pedagogical rewards of having

to grapple with the challenge of thinking and writing strategically. We deprive them of the benefits of their own creativity, the fruit of trying labor. Let us recall that the 1930s generation of students at the U.S. Army War College produced the Rainbow Plans for World War II. According to Henry Gole, a distinguished instructor at the U.S. Army War College in the 1990s: "The work produced by the students, staff, and faculty beginning in 1934 at the Army War College anticipated the very conditions faced by the United States in 1939-41."[19] In those simpler days with of a much smaller military establishment and a nascent national security decisionmaking process, Major (later General) Albert C. Wedemeyer in 1941 wrote the victory plan for World War ll.[20] To improve competence in strategy, students should write papers on grand strategy and military strategy. Accordingly, they would learn the value of connectivity and constant two-way feedback between the higher and lower realms of strategy, from grand strategy to military strategy, as well as the integration of the instruments of national power with military strategy.

4. Mine extensively the case study method so that students understand how to make strategy. Case studies are the most effective tools for adult learning, they force students to become intellectually engaged in confronting the dilemmas of decisionmakers. In-depth case studies should be interwoven throughout the curriculum, not simply appended here and there, so that students fathom the correlation of theory with facts, and with the cognitive dimensions of the strategist. The success of the Vietnam and the NSC 68 cases testify to the pedagogical value of case studies. Possibilities abound: the decision to go to war, conflict termination, and post-conflict reconstruction and stab-

ilization. There are some case studies available from Harvard and Georgetown, but they do not address gaps in strategic pedagogy. We should develop our own, tailored to the learning objectives we want to achieve, such as the appropriate strategies for the levels of war. Case studies should demonstrate the integration of national security strategy and military strategy, at all levels in the spectrum of conflict and phases of war, in addition to the instruments of national power. The case studies should contain a balance of ambiguity, competing global and regional priorities, moral hazard, insufficient intelligence, policy contradiction, and competing demands for resources. The human mind learns more from frustration than from success. Success merely confirms predispositions and sets up the unpleasant prospect of the strategic surprise of failure.

5. Emphasize total strategy, the integration of the instruments of national power, in regional studies courses. Competence in strategy requires a sophisticated understanding of the state's and society's sources of power, strategic culture, and the employment of national and international resources to achieve the ends of policy. Since the United States is a global power with regional security responsibilities across the spectrum of conflict, students need to have some understanding of how to work effectively in foreign cultures, as well as how to balance priorities within competing global, regional, functional requirements, and the interagency dimensions of these responsibilities. Regional studies are a superb vehicle for teaching about how the interagency works, of bringing to bear the kinetic and nonkinetic, whole of government elements of power. At the same time, understanding the interagency synergy adds

immensely to the kit bag of the budding strategist by acquainting him or her with the culture and capabilities of other departments and agencies. Regional studies, along with case studies, are the best way to study and learn total strategy of the kind contained in NSC 68, the kind required by today's complex unconventional challenges to national security.

6. Deal with real world strategic issues by employing living (with an appropriate nod to retaining historical) case studies. This would beef up teaching all of the components of the Policy-Strategy-Interagency-Instruments of Power-Culture curriculum continuum. Such an approach has another virtue: students would become acquainted with the organization of the policy and strategy communities, the true interagency in Washington and in the field, from the President, to the National Security Council, Department of State, DoD, Homeland Security, embassies, combatant commands, and allies. The following regional case studies are examples of what can be done for delivering the above continuum well:

Latin America—Colombia, state building, rule of law, counternarcotics, counterterrorism.

Europe—North Atlantic Treaty Organization (NATO), coalition strategy, out of area operations.

Middle East—Iraq, high intensity warfare, conflict termination, post-conflict reconstruction and state building, nuclear proliferation.

Africa—humanitarian crisis, environmental secur-ity, failed states, diseases.

Russia, Commonwealth of Independent States (CIS)—European and Middle East security, state building, partnership for peace.

Asia—multiplicity of power centers, peer competitors, Afghanistan, India-China, nuclear proliferation.

7. As the content and pace of courses changes to emphasize problem solving, emphasize the importance of writing strategy effectively. This would require that students have more time to analyze and write. Of all the forms of learning, writing is second only to actual experience. As mentioned above, the problem solving tasking should be introduced early in the curriculum and completed at logical intervals along the way. For example, students could be tasked to develop strategy for war termination and post-conflict reconstruction. The intellectual reward consists of evaluating and applying the gamut of strategic principles, from realism to idealism, the center of gravity, just war, war as policy by other means, the integration of the instruments of power, and many more.

8. Modify the calendar in order to allow maximum time for faculty and student preparation for problem solving learning. A crowded course schedule suboptimizes faculty and student learning because of quick turnarounds, multiplicity of requirements, and competing nonacademic requirements. Curriculum planners need to perform surgical triage and choose the most pedagogically rich academic events, lesson plans, case studies, and exercises. They need to sweep away passive sitting time and move to truly active learning.

9. Establish in conjunction with a university a doctoral program in strategy. Currently, no American university has such program and doctoral studies in military history are becoming scarce. One of the best doctoral programs of this nature is the war studies program at the Royal Military College of Canada in Kingston, Ontario. It is government funded, open to military officers and civilians, and has an excellent reputation for teaching and research excellence.

In summer 2009, the U.S. Army War College was exploring the feasibility while the Air War College had already established a doctoral program.

10. The last recommendation may be the most challenging: modifying the culture of the war colleges. The seminar-centric model of pedagogy has great rewards. It promotes bonding and mutual learning, qualities essential to effective military organizations. Interactive learning can bring out the best among seminar mates. But the seminar may not be the best mode for learning strategy. The war colleges should rebalance the seminar-based pedagogy with scheduled time for individual study. This would bring them closer to the academic culture of a graduate institution.

These are potentially revolutionary initiatives. Implementing them will require a different approach to the curriculum and a different form of faculty preparation, because the pedagogical emphasis would be on analyzing problems and developing strategy while maintaining a foundation in theory. The new approach demands pruning the curriculum overgrowth and focusing on pedagogy that has the highest payoff. The war colleges are great institutions whose potential we have not fully tapped. We need to retire old approaches gracefully, move forward creatively, and become the nation's preeminent centers for teaching strategy.

ENDNOTES - CHAPTER 4

1. An earlier version of this chapter appeared in the January 2009 issue of *Joint Force Quarterly*.

2. A few military students selected for military education level 1 (war college) do fellowships either at a university, a think tank, or a department/agency of the federal government, in lieu of

the program at Carlisle, Newport, Montgomery, Washington, or Quantico. A smaller cohort of military students frequents foreign war colleges. Selection for attending a war college is highly competitive and is done by a process external to the colleges, a distinction from civilian universities, which pick their students from a pool of applicants.

3. Since the late 1990s, all war colleges have been required to grant a master's degree, which involved a process of accreditation by the Department of Education, the Joint Chiefs of Staff, and the regional accreditation bodies. In addition, the House of Representatives Committee on Armed Services exercises jurisdiction over the curriculum through visits and inquiries.

4. The HASC Subcommittee on Oversight and Investigations held a series of hearings in 2009, engaging the testimony of scholars and war college and command and staff college commandants on the matter of whether the schools are doing a good job in educating strategic leaders. At the May 20, 2009, hearing, subcommittee chairman Vic Snyder stated:

> From time to time we should assess what our professional military schools are meant to do for the nation. We are going to ask if they are doing what the nation needs now and whether they are doing it in the best way. . . . We are going to be asking about the mission focus of each school; whether they pursue their program with vigor; what they are doing to ensure the highest quality staff, faculty, and students; and whether the curricula has the appropriate balance between the enduring elements of what military professionals need to know and what they need to understand about the new challenges facing the nation.

See: "Oversight and Investigations Subcommittee Holds Hearing on Professional Military Education," available from *www.house. gov/list/press/armedsvc_dem/OIpr052009.shtml*.

5. See Alexander S. Cochran, Testimony to House Armed Services Subcommittee on Oversight and Investigation, "20 Years Later: Professional Military Education," May 20, 2009. Cochran

warns about protectiveness: "In summary, while all war colleges are justifiably proud of their programs, my experience suggests that this pride has created intense protectiveness on their own prerogatives on both the process and substance of their operation." p. 4.

6. Richard Betts, "Is Strategy an Illusion? " *International Security*, Vol. 25, No. 2, Fall 2000, p. 5.

7. *Ibid.*, pp. 6-7.

8. Beaufre, *Introduction a la Strategie*, Paris, France: Economica, 1985, p. 16. This definition approximates Clausewitz's notion that

> war is nothing but a duel on a larger scale . . . if you want to overcome your enemy you must match your effort against his power of resistance, which can be expressed as the product of two inseparable factors, viz. the total means at his disposal and the strength of his will. The extent of the means at his disposal is a matter — though not exclusively — of figures, and should be measurable. But the strength of his will is much less easy to determine and can only be gauged approximately by the strength of the motive animating it.

Carl von Clausewitz, *On War*, Michael Howard and Peter Paret, eds. and trans., Princeton, NJ: Princeton University Press, 1976, pp. 75, 77. Beaufre was deeply influenced by experience in World War II, the French in Vietnam and Algeria, the advent of nuclear strategy, and the NATO alliance. Beaufre's book was published in English (with a preface by B. H. Liddell Hart), *An Introduction to Strategy, With Particular Reference to Problems of Defense, Politics, Economics, and Diplomacy in the Nuclear Age*, New York: Praeger, 1965.

9. An excellent guide to making policy and its relationship to strategy is Richard L. Kugler, *Policy Analysis in National Security Affairs: New Methods for a New Era*, Washington, DC: National Defense University, 2006.

10. David Jablonsky, "Why Is Strategy Difficult?" in Boone Bartholomees, ed., *US Army War College Guide to National Security Issues, Vol. I: Theory of War and Strategy,* Carlisle, PA: Strategic Studies Institute, 2008, p.3.

11. *Ibid.,* p. 10. According to alchemy, the philosopher's stone would turn base metal into gold.

12. To be fair to the historical record and to the difficulty of the art of strategy, they also made mistakes, Pericles and Churchill notably.

13. See the writings of Samuel Huntington, Russell Weigley, Antulio Echevarria, Colin Gray, and Nigel Aylwin-Foster on the American way of war and strategy.

14. The volume edited by Williamson Murray, MacGregor Knox, and Alvin Bernstein, *The Making of Strategy: Rulers, States, and War,* New York: Cambridge University Press, 1994, is an excellent comparative study of how nations and statesmen approached the making of grand strategy.

15. Colin Gray, *Irregular Enemies and the Essence of Strategy: Can the American Way of War Adapt?,* Carlisle, PA: Strategic Studies Institute, U.S. Army War College, March 2006, p. 6.

16. The following matrix of questions is condensed from Alan G. Whittaker, Frederick C. Smith, and Ambassador Elizabeth McKune, "The National Security Process: The National Security Council and the Interagency System," Washington, DC: National Defense University, September 2004, pp. 24-26.

17. George E. Hein, "Constructivist Learning Theory," Institute for Inquiry, 2007, available from *www.exploratorium-edu/ifi/resources/constructivistlearning.html.*

18. Steve Fought, "The War College Experience," *Academic Exchange Quarterly,* June 24, 2004, pp. 1-2, available from *www.thefreelibrary.com/The+war+college+experience-0121714082.*

19. Henry Gole, *The Road to Rainbow: Army Planning for Global War, 1934-1940*, Annapolis, MD: Naval Institute Press, 2003, p. 31.

20. Charles E. Kirkpatrick, *An Unknown Future and a Doubtful Present: Writing the Victory Plan of 1941*, Washington, DC, Center of Military History, 1992.

CHAPTER 5

TEACHING STRATEGY:
A SCENIC VIEW FROM NEWPORT

Bradford A. Lee

These days it is hard to find a pleasing scenic view of strategy. Dark clouds of pessimism have descended upon experienced professional observers of American strategic performance. Andrew Krepinevich and Barry Watts, unusually sharp and well-informed analysts in a first-rate think tank in Washington, DC, have recently been arguing that the American government has been unable for several decades to do strategy competently. Perhaps even more depressingly, they cast doubt on the ability of institutions of professional military education to teach military officers and government officials how to do better. In their view, most students of strategy simply do not have an intellectual knack for the subject.[1]

Lest one suppose that such views merely reflect dismay over American strategic performance in Iraq and Afghanistan, one should recall that even before the September 11, 2001 (9/11) attacks, Richard Betts, an accomplished senior statesman of strategic studies in civilian academia, suggested that no government, especially in a modern bureaucratic state, could reasonably expect to use strategy for its fundamental policy purpose: to turn political objectives at the start of a war into desired political outcomes at the end of the war. While granting that strategy is necessary in principle to make war rise above mindless murder, Betts presented a formidable array of obstacles that in his view render strategy almost impossible in practice—in the sense that strategic concepts for connecting military operations and government policy can actually be a "value added" factor, "a cause of victory that can be distinguished from raw power."[2]

Somewhat sheepishly, I must confess to making my own contribution to the *Zeitgeist* of strategic pessimism. On the basis of research that I have done on each American war from 1917 to 2001, I concluded in 2002-03 (on the eve of Operation IRAQI FREEDOM [OIF]) that critics were mostly on the mark in pointing out the United States wins wars but loses the peace. To be sure, except in the Vietnam War, the United States has been able to use military force to achieve its most basic layer of political objectives, in the sense of determining who should have, or not have, control over the central chunk of territory being contested. But in all of its wars since 1917, American policy has had loftier levels of political purpose than that. Notwithstanding the operational prowess that American military forces have usually displayed, political outcomes in those higher layers of purpose have either been much less favorable or much less durable than the policymakers who took the United States into each war anticipated.[3]

If economics is the dismal science, strategy must be the dismal art. But since no one can reasonably decide on a war, effectively fight a war, or successfully seek to prevent or end a war without having in mind some sort of strategic assumptions about what actions will have what consequences in the real world, the study of strategy is something that practitioners can avoid only if they have no concern for improving their odds of success.[4] Having spent most of my adult life teaching future practitioners about strategy, I can attest to the intuition of Krepinevich and Watts that many military officers and government officials do not find that thinking strategically comes easily or naturally to their minds. There is enough truth in the case for strategic pessimism that those who teach practitioners ought to be humble and hard-working. It is not, however, the whole truth. Formal education can help some minds to work in more strategically effective ways and other minds at least to stop working in strategically self-defeating ways.

An essay on teaching a strategy course needs a strategy of its own. We can think about the Strategy and Policy (S&P) course of the U.S. Naval War College in terms of its practical purposes, the human capital accumulated in its faculty, and the learning aids that the faculty lays out for its students to use. In that sequence I shall make my scenic excursion of how my colleagues and I teach strategy at Newport, Rhode Island.

PRACTICAL PURPOSES

Rear Admiral Jacob Shuford, recently retired after 5 years as president of the Naval War College, has eloquently made the case for "humanistic re-education" of 21st-century warriors.[5] His successor in Newport, Rear Admiral Phil Wisecup, is so committed to the worth of such an education for senior officers that he was engaged intellectually with members of the Department of Strategy and Policy even while he was still in command of the *Ronald Reagan* strike group launching sorties into Afghanistan. Both follow in the tradition of an earlier Naval War College president, Stansfield Turner, who in the 1970s (another time of strategic despair) restored the broad study of historical experience, added the close reading of great books written in the past about war and strategy, and also instituted in the Strategy and Policy department the tutorial system of liberal education that he had encountered as a Rhodes Scholar at Oxford. These liberal-arts traditions remain alive and well in our curriculum, and many officers who take our course value highly the opportunity to read and discuss what the great minds of the past have to offer about timeless issues very much on the minds of contemporary warriors.[6]

As Turner affirmed, however, at the Naval War College we do not engage in such learning for its own sake.[7] We are a professional school, not a liberal-arts refuge from the real world. Our primary practical purpose is the professional development of officers and officials who can make wise strategic choices and resolve tough strategic problems. We are preparing them for senior leadership in positions where the decisions they make will have strategic consequences, for better or for worse. If they are to be better, high-level commanders and high-level staff officers need powers of judgment that draw on a broad perspective enabling them to see beyond the here-and-now, and they also need tools of analysis that dig deeper than the conventional wisdom, the doctrinal dogmas, and the buzzword blizzard swirling around them.

It is not sufficient for them to figure out the best course of action in a given situation. It is also necessary for them to convince others of the wisdom of what they propose to do. Otherwise their ideas will be stillborn in the real world. So our secondary purpose is the development of powers of persuasion. That means military officers (and civilian officials) who can communicate effectively, with both the spoken word and the written word. We give our students plenty of educational opportunity to do so, without allowing them to exercise intellectual force protection with the bullets of PowerPoint slides or one-page point papers. They are assigned a series of three 10-page essays in which they must not only develop an argument but also confront counterarguments (with the latter looming as a source of dread for many students). Every week they have to discuss and debate issues with mental and verbal agility in lengthy but fast-paced seminars (with one-quarter of their overall grade at stake to help loosen tongues). Before each of their first two essays, students take on their faculty co-moderators, one on two, in a tutorial, a verbal sparring match of 30 minutes or more (the intensity of which can be another source of dread for some students). For their third essay, which serves as a final exam, they are entirely on their own, forced to fuse the various intellectual elements of the course in a coherent analysis of some issue of enduring relevance or contemporary urgency. In these educational exercises of reasoned persuasion, smart officers come to appreciate how different types of argument might resonate, or not, with various audiences: political leaders, military leaders, interagency counterparts, foreign coalition partners, international organizations, people caught between insurgents and counterinsurgents, and the general public at home and abroad. Just as we teach that there must be a match between strategy and policy, so we also teach that there should be a match between arguments and audiences.

HUMAN CAPITAL

The Strategy and Policy faculty is distinctive in the U.S. system of joint professional military education (JPME) in two ways worth noting. In our seminars, the crucible of our course, we always teach in two-person moderator teams; for each group of students, a civilian academic is paired with an active-duty military officer. And the faculty teaches (as do the other two teaching departments at Newport) both the Intermediate Level Course (ILC) and the Senior Level Course (SLC) of JPME each year.

The ILC, called Strategy and War, is tilted toward the interface between strategy and operations. For this course, paraphrasing Clausewitz, we define strategy as the use of operations for the political purpose of the war under consideration. Each week we consider a different war, with each new one having different political purposes. In the Clausewitzian tradition, we constantly remind the majors and lieutenant commanders that operations along with strategy must be aligned with policy. The SLC, called Strategy and Policy, not only gives more attention than the ILC to nonmilitary instruments of power and nonmilitary dimensions of war, but also features individual case studies that span more than one type of war and stretch across several decades, so that we can fully incorporate prewar, postwar, and interwar periods in our course, even as war itself remains the central focus. In the SLC, our definition of strategy becomes the use of war for the purpose of long-term political success in the international arena, and our perspective on policy becomes quite complex.

Viewed from either of these two perspectives, strategy is the bridge between policy and operations, and accordingly, it makes great sense to combine in each seminar-moderator team a professor with a civilian academic background oriented toward policy and an officer with professional experience oriented toward operations. A positive by-product of this joining together of two types of human capital is that a good team of such co-moderators can provide an example of effective collaboration across the civil-military divide. (In my aforementioned analysis of problems in U.S. war-termination strategies, I high-lighted the dysfunctional lack of close and frank communication between civilian leaders and military leaders.[8]) A further benefit is that we can expeditiously handle turnover in our faculty roster by teaming a new military moderator with a seasoned civilian moderator, or vice versa. That teaming allows us to put relatively inexperienced teachers in the classroom right away and have them learn by doing with their more experienced co-moderator. Most new military faculty members are more likely than most new civilian faculty members to feel comfortable with the ILC; the intellectual comfort zones are likely to be reversed for the senior course. As a team, the civilian academic and the military officer can complement each other over the two courses. We preach the virtues of versatility and adaptability to our students; we practice those virtues as a faculty going back and forth between the ILC and the SLC.

The SLC as it currently stands is perhaps the most intellectually sophisticated and challenging course ever taught in an institution of professional military education. Students as well as faculty have risen to the challenge. The first time that we taught the current version in 2007 (beginning only 1 month after the end of the intermediate course), the results were gratifying. On a scale of 1 to 7 (with 7 being the most positive response), a survey of students at the end of the term generated the following statistical results:

Degree you were satisfied as a student in S&P course:
6.24

How would you rate yourself as a student:
5.38

Hours per week spent on required reading:
37.2

How well faculty moderators facilitated discussion in seminar:
6.20

Benefit of civ/mil teaching team vice single moderator:
6.24

Percentage of lectures rated 5.75 or higher
74.5%

Note the amount of time devoted to the weekly reading assignments of 600 pages or so. To the extent that interest in, and commitment to, learning about strategy eventually translates into proficiency in doing strategy, the apprehensions of Krepinevich and Watts may be overblown.

Maintaining a high educational standard requires a first-rate faculty as well as a committed student body. The long-term trend in faculty recruitment and retention is fraught with uncertainty. In principle, the ideal military faculty member has had both command and combat experience and has not only completed our course with distinction but has also done advanced graduate work (perhaps Ph.D. work) at a top civilian university. In practice, especially given the operational tempo of the post-9/11 era, that ideal is hard to attain. Over the years, it is our Army faculty members who have come closest to it. Recently, we have had a strong Air Force contingent as well. But Army and Air Force officers sometimes do not stay with us for a full 3-year tour. In the pre-9/11 era, Marines punched above their weight in numbers on the Strategy and Policy faculty. Since 9/11 "the few and the proud" have become even fewer among us (though, of course, no less proud). The Navy personnel system does not come close to filling all faculty billets designated for naval officers. If the Long War continues to live up to what was once its Department of Defense (DoD) name, intense patterns of deployment of officers will take a cumulative toll on the strength of our military faculty. Already, in the past few years, several faculty members have been pulled out of our classrooms and put into Iraq or onto staffs in the Persian Gulf region. That has wreaked havoc on our teaching teams and required quick adaptation on our part.

Recruitment and retention of civilian academics also require adaptability. When I arrived at Newport in 1987, most civilian faculty members in the Strategy and Policy department were sojourners. They would come for a year or two and then return to civilian universities or go on to policymaking positions in the government. On the one hand, this in-and-out pattern had some positive spin-off effects. Our faculty alumni set up courses modeled on ours at top universities: Yale, Harvard, Johns Hopkins (School of Advanced International Studies), University of Pennsylvania, and elsewhere. Our faculty alumni have also had influence on policy and strategy inside the beltway throughout the past 2 decades. On the other hand, the in-and-out pattern of recruitment and retention kept the Strategy and Policy curriculum from reaching its full potential. Thus, in the 1990s we sought to build the department on a stronger foundation of long-term civilian faculty members who could drive a process of continuous and cumulative improvement in the sophistication and relevance of our core course. Our recruitment strategy took advantage of flaws in the civilian academic market. We were able to hire excellent mid-career scholars in the fields of diplomatic and military history or of security studies and international relations who were denied tenure or lacked attractive job opportunities at top universities.

We have since lost some of these outstanding professors to death, retirement, government service, or more lucrative and prestigious academic positions elsewhere. We have gained first-rate regional specialists (more than one-quarter of our civilian faculty has had substantial training in East Asian languages). But we suffer from the fact that, except for our faculty alumni, few professors in the civilian academic world teach strategy as we do. The theoretical involution of the political-science discipline, the tendency of its security-studies and international-relations sub-disciplines to skirt "around the edges of war" (in the words of my colleague Tim Hoyt), the ideological aversion of history departments to military and diplomatic historians, the long-term decline in intellectual talent going into the academic discipline of history in the United States, and the intense post-9/11 competition for linguistically proficient specialists on the Middle East and South Asia (those precious few willing to work for the U.S. Government) have all made it hard for us in recent years to recruit mid-career civilian professors who have both the educational background and the intellectual adaptability to add immediate, significant value to the Strategy and Policy curriculum. Changing our recruitment strategy yet again, we have shifted our focus to younger scholars who have just finished their doctoral dissertations but have not yet fallen irretrievably into disciplinary ruts. Just as younger military officers seem to have adapted more expeditiously than many of their seniors to unanticipated forms of warfare in Iraq and Afghanistan, so we expect our younger recruits to adapt quickly to the distinctive ways in which we approach the study of strategy in Newport.

LEARNING AIDS

Like all JPME departments, Strategy and Policy has many sources of guidance about the substance of our curriculum. Congressman Ike Skelton and his colleagues on Capitol Hill are one such source. The Joint Staff, with the various learning areas and objectives of its Officer Professional Military Education Policy (OPMEP), is another source. Presidents or provosts at the Naval War College sometimes suggest points of emphasis. We also pay attention to the general declarations of strategy that emerge from the White House, the interagency process, the Office of the Secretary of Defense, the Joint Chiefs of Staff, and — not least — the Navy itself, which unveiled a new Maritime Strategy at the International Seapower Symposium in our lecture hall in 2007. In truth, many of these "strategy" documents do not actually represent strategies in the fundamental sense that we teach about in the Strategy and Policy department. Instead they are often pious bureaucratic expressions of other-worldly aspirations. Still, it is important that our students know about these aspirations, and so these documentary leaps into the stratosphere do have influence on how we present our course.

Ultimately, however, it is the Strategy and Policy faculty that determines the ways in which we educate officers and officials in the art of making and executing strategy in support of policy. Throughout my twenty-two years in Newport the leadership of the Naval War College has faithfully upheld that vital principle of academic freedom. For all the guidance and suggestions that we get, there have been no military orders from above that have denied us the ability to shape and reshape the curriculum as we see fit. Over an extended period we have been able to hone the three key "learning aids" that together constitute our distinctive approach to the teaching of strategy: (1) course themes; (2) classical theorists; and (3) a mixture of ancient, modern, and contemporary case studies.

COURSE THEMES

When officers move into our syllabus, the first landmark that they encounter is the course themes. A rudimentary version of those themes dates back to the Turner reforms of the 1970s at the Naval War College. Skeptical of attempts by military officers (including Alfred Thayer Mahan) to ransack history for immutable principles of war, and chary of efforts by educators to coach students toward school solutions, the chairman of the Strategy and Policy department at that time, Philip Crowl, affirmed that past experience can only help us ask the right questions about a current or future war, questions that define the problems we shall face in any war. In a Harmon Memorial Lecture at the U.S. Air Force Academy in 1978, Crowl laid out six such questions, or sets of questions.[9] More than 3 decades later, the format of questions remains, but the number of questions has proliferated mightily and now fall under 11 headings for the SLC (and nine for the ILC).

The first seven SLC headings we group together as "Matching Strategy and Policy: The Process."
1. The interrelationship of policy, strategy, and operations,
2. The decision for war,
3. Intelligence, assessment, and plans,
4. The instruments of national power,
5. Interaction, adaptation, and reassessment,
6. War termination,
7. Winning the peace and preparing for war.

The last four headings come together as "Matching Strategy and Policy: The Environment."
8. The international dimension of strategy,
9. The material dimension of strategy,
10. The institutional dimension of strategy,
11. The cultural and social dimensions of strategy.

The process themes invite future strategic leaders to think through entire cycles of war and peace playing out over time in a given case, while the environment themes remind them that how appropriately strategies are chosen, and how effectively they are executed depends crucially on the multiple contexts in which strategic leaders operate. If we were to add yet another theme in the future, it would no doubt be about the information and communications environment, but as the themes stand now, issues relating to information operations and strategic communication receive substantial coverage under themes 3, 4, and 11. Issues relating to technology find their place in themes 4, 7, and 9. All themes consist of three full paragraphs of questions, except for theme 4, which requires five paragraphs to cover diplomatic, information, military, and economic instruments (DIME) and jointness.

The first course theme stands out above all, because of the vital importance of ensuring that military actions (and nonmilitary actions as well) are closely tailored to the political purposes of the war at hand. The first question in that theme points officers and officials toward the realm of policy: "What were the most important political interests and objectives of the antagonists?" Both the SLC and the ILC plunge students into the complexity of policy, especially American policy. They come to see that policy may have different layers to it. One layer in all the wars that we study has to do with political control over contested territory. That control may involve either limited or unlimited (regime change) political objectives with respect to who controls what territory. A second layer, often a playground for great powers, has to do with the balance of power in a region of conflict. One side may seek to defend or restore what it perceives as equilibrium; the other side may seek to tilt the balance in its favor or, more ambitiously, dominate the region. A third layer, often in the political vision of ideologically minded strategic actors, has to do with norms or values governing the international system. One side may seek to vindicate existing principles; the other side may seek some sort of new world order.

When officers get to the second half of the SLC, in which the U.S. experience as a global power looms large, they see that American political leaders, and also political leaders of the primary U.S. adversaries since 1941 (Nazi Germany, the Soviet Union, and al Qaeda), usually have had political purposes extending through all three layers. Matching strategy to multilayered policy is no simple affair. It becomes even harder when policy guidance is not very concrete in laying out political objectives, but is much more concrete in setting political restraints on what military strategists may choose by way of courses of action, as is typically the case when the United States is at war.

The SLC is about victory in war. Examining the political outcomes of American wars in this context, officers in the course get a deeper (but perhaps unhappier) sense of the complexity of the concept of victory. In using military force in a given case, the United States may achieve different degrees of political success in different layers, degrees of success that may play out over different time horizons and with different balances of cost and benefit. To envision a singular and static political endstate may well provide a useful focus for strategic and operational planning — but one unlikely to crystallize once and for all in the real world of dynamic interaction between willful adversaries.

CLASSICAL THEORISTS

As officers move forward in the syllabus from the course themes to the weekly modules of SLC, the first module thrusts them deep into the world of theory. Rather than expose them to bite-sized excerpts from or canned summaries of many different theorists of war, contemporary as well as classical, the Strategy and Policy faculty concentrates their minds first and foremost on the two most venerable and most valuable texts of strategic theory: Clausewitz's *On War* and Sun Tzu's *Art of War*.[10] We do not approach these texts as an intellectual historian would. To be sure, a lecture by one of our China specialists, Dex Wilson, and an introductory commentary by Samuel Griffith (USMC and Oxford Ph.D.) provide for our students some of the ancient Chinese cultural and political context of the Sun Tzu text, while a later case study takes them through the era in which Clausewitz fought and wrote. But we do not spend much time trying to figure out what the two great classical theorists *really* meant by their cryptic but celebrated utterances. (Having read some of the two texts in their original language, I can testify that it is a fool's errand for nonlinguists to make pronouncements about the real meaning on the basis of modern translations.[11]) For the most part, we make pragmatic forward-looking use of the texts, rather than let the texts make backward-looking philological use of us.

We expropriate from Clausewitz his most important method for educating strategic leaders: *Kritik* ("critical analysis").[12] This method turns our students toward counterfactual analysis of alternative strategies as we examine cases of actual strategies that did not deliver the desired political outcome. Such a counterfactual twist would be a wrenching dislocation for most historians in the academic world, but it is an important vicarious exercise for future decisionmakers in the real world. Guided by Clausewitz's exposition of his method, officers think through chains of cause and effect that might connect military means and political ends; come to a more mature appreciation that what makes sense at a lower level of war might not make sense at a higher level of war; see how "[t]he higher the ends, the greater the number of means by which they may be reached"; and take on board his realistic view that "[i]n war, . . . all action is aimed at probable rather than certain success."[13]

We also extract from Clausewitz some basic substantive ideas and important descriptive concepts. His emphasis on war as a continuation of policy "with the addition of other means"[14] is the inspiration for our first course theme. His admonition that both statesmen

120

and commanders should try hard to anticipate the nature of the war on which they are embarking informs a central point of our third course theme and influences the way in which we sort out the types of wars that we cover in our case studies. His description of war as the realm of fog, friction, uncertainty, and chance remains quite valid in the 21st century, as many officers at war colleges have already learned from personal combat experience. As our course proceeds, we often hearken back to Clausewitz's likening of war to a trinity in a paragraph on the last page of his first chapter, and we also make frequent use of his commentary in the next paragraph on the roles of government, military, and people — which we relate as the Clausewitzian triangle, to distinguish institutional elements from the rational, creative, and emotional aspects of the trinity.

Clausewitz is more ambivalent about prescriptive concepts than almost any other military theorist, and in the Strategy and Policy course we are ambivalent about his quasi-prescriptions that have smitten doctrine writers. We note ambiguities in and limits to the application of his center of gravity concept at the strategic level of different types of war and invite officers to ponder what alternative mediating concept between military action and political outcome might work better in a variety of particular cases.[15] We note that Clausewitz is more attached to concentration (a.k.a. mass) than to any other principle of war, but we also note that he is somewhat circumspect about applying this principle to multitheater wars — wars that figure prominently in our case studies.[16] In discussing our course theme on war termination, we note as well the tension between what Clausewitz says about not overshooting the culminating point of victory and what Michael Handel has identified as Clausewitz's "principle of continuity."[17]

As an intellectual counterbalance to Clausewitz, Strategy and Policy faculty also point students toward Sun Tzu's prescriptive notions.[18] We seek to use this old Chinese text to generate new ideas. We turn Sun Tzu's exhortation to know your enemy and know yourself into an elaborate process of net assessment — the outstanding historical example of which comes from another old book, by Thucydides (discussed below), and an interesting American example of which is George Kennan's famous "X" article of 1947.[19] When Sun Tzu prescribes that the best strategy is to attack the enemy's strategy (what Kennan was in effect prescribing) and that second best is to attack the enemy's alliance, we stretch officers' creative capacities by having them think through what those two leading items on Sun Tzu's menu of options might mean against adversaries in types of conflicts absent from his text — not just in a cold war, but also in an insurgency of either the Maoist or the al Qaeda and Associated Movements (AQAM) variety. In a more operational vein, what Samuel Griffith translates as "strategist's keys to victory" (which I dub "interaction games") include ways to outmaneuver an enemy that are quite different from the traditional American way of war and might be useful against irregular as well as regular forces. The notion of hybrid warfare, so much discussed nowadays, we find in Sun Tzu's discussion of *cheng and ch'i* forces.[20] Putting the enemy on the horns of a dilemma can be something that we do to our enemies, not just what our enemies do to us.

The two classical theorists stand out above all other, more recent, theorists in how they address what I call the two mega-concepts of strategy: rationality and interaction. Each of those key concepts has two faces to it. Clausewitz looks carefully at means-ends rationality and even more closely at cost-benefit rationality. Costs in turn depend on the dynamics of interaction with the enemy. He plays up how difficult it is to anticipate enemy military reactions to one's own military actions, while perhaps not giving enough attention to the other face of interaction — how the enemy will react politically to what one does. No theorist of war has superseded what Clausewitz had to say about the mega-concepts.

Sun Tzu gives hyper-rational elevation to the first mega-concept by highlighting the idea of gaining as much political benefit as possible at the smallest possible cost in military force. In a hypothetical dialogue between the two classical theorists (imagined in a scintillating lecture given by my colleague, Karl Walling), Clausewitz would argue that the escalatory dynamic of interaction makes Sun Tzuian hyper-rationality a pipe dream. Sun Tzu's counterargument would be that he offers the keys to mastering interaction. In the terminology of our era, those keys are information superiority and maneuver warfare. Remarkably, the ghosts of this hypothetical debate between Clausewitz and Sun Tzu have lurked in the actual debate between enthusiasts and skeptics of transformation in the American defense community at the turn of the 21st century.

Other strategic theorists can be evaluated with regard to the mega-concepts. Not surprisingly, at the Naval War College, we bring Mahan and Corbett into the Strategy and Policy course not long after we introduce Clausewitz and Sun Tzu. From those two classical naval thinkers, we derive the three naval warfighting missions of greatest strategic importance: fleet-on-fleet engagement to achieve command of the sea; power projection from the sea onto the ground; and economic/logistical warfare via the interdiction of sea lines of communication. Mahan posited that good naval commanders must be great risk takers with their fleet, intent on destroying or neutralizing the enemy's fleet. When and where to risk the fleet is indeed the most strategically important decision that a naval commander must make. We invite officers in seminar discussion and in essays to ponder whether Mahan, committed as he was to the operational principles of concentrating forces and taking the offensive, was sufficiently rational in weighing risks and rewards of fleet-on-fleet engagements in the context of pertinent strategic circumstances.

Corbett, for his part, was interested in a wider range of naval operations than Mahan seemed to be, but we are particularly interested in what this British theorist suggested about projecting ground forces from the sea. More concerned with risk management than Mahan evidently was, he played up joint operations in peripheral theaters, where sea control could be exercised, as a relatively low-risk way to achieve a major strategic payoff in an ongoing war. We ask officers to evaluate whether or not Corbett measures up well against the means-ends aspect of the rationality mega-concept. Can sideshows of maritime strategy really make a major contribution to winning wars? If so, under what circumstances can this happen? Our judgment of the general rationality of both Mahanian and Corbettian theories turns out to depend heavily on the judgment of particular strategic circumstances. But perhaps that is true of all theory; perhaps no theory can guide practice in all circumstances.

The Strategy and Policy department currently does not require students to read anything written by air power theorists, though we have assigned the work of Giulio Douhet and John Warden in past years and may do so again in the future. Like Sun Tzu, air power theorists incline toward hyper-rationality; but unlike Sun Tzu, they have a debilitating propensity simply to ignore the interaction mega-concept. They also are wont to focus single-mindedly on one warfighting mission of the air instrument—strategic attack (a.k.a. strategic bombing)—at the expense of missions in which it operates more jointly with ground forces. Even with regard to strategic attack, there is much room for progress (indicators of which can be found in recent historical work on World War II) in our understanding of the strategic effects of bombing on the economic and political systems of different types of enemies. The most pressing need, however, is for a theorist to plumb the underlying interaction dynamics of how the United States, in its regional wars from 1990 to 2003, has been able to generate such major asymmetric advantages at a remarkably low cost, while fighting the *regular* forces of its adversaries, by the way that it has used air forces and ground forces in various combinations. The follow-on theoretical inquiry would then extend to the underlying interaction dynamics of why from a strategic perspective there has not been such a one-sided advantage against enemy *irregular* forces. More generally, if the primary mandate of JPME institutions is to highlight jointness, all of us teaching strategy in those institutions could benefit greatly from a first-rate theory of the strategic payoff, in different types of wars, of joint operations or, more broadly conceived, the integration of different instruments of power and forms of influence. The SLC case studies provide historical raw material for such a theoretical product.

CASE STUDIES

It is case studies, the third learning aid of the Strategy and Policy course, which take up the bulk of the syllabus. Between the first week of the SLC, when we introduce the classical theorists, and the final week of the SLC, when we explore possible futures of maritime strategy, officers come to grips with 10 case studies.[21] Our earliest case study takes students back almost 2 1/2 millennia to the Peloponnesian War. The next six case studies span 2 centuries, from the 1770s to the 1970s. The final three case studies deal with ongoing strategic problems for the United States and its coalition partners: wars with and within Iraq 1990-2010; the wars that have threatened to erupt over the proliferation of nuclear weapons, from China, to India and Pakistan, to North Korea, and on to Iran; and the global war against revolutionary jihadists, which we trace back over 3 decades and follow up to the current year. Our shortest case in chronological scope covers 11 years, 1940-51, as we investigate the policy and strategy issues involved in the U.S. rise as a global power from the early months of World War II to the middle of the Korean War. Our longest case runs over half a century, as we examine the strategic ups and downs of Britain from the end of the Seven Years War in 1763 to the end of the Napoleonic Wars and the establishment of a Pax Britannica based on command of the maritime commons.

Traversing such vast sweeps of history is not an end in itself. Rather it is simply one of the ways by which we seek to develop strategic leaders with broad perspective on the problems that they will face in the future. We certainly do not expect our students to master much historical detail, though some officers do come out of our course as enthusiastic history buffs.[22] We do expect our students to grasp the big picture of each case, discern its patterns and puzzling turning points, and build intellectual bridges to other cases. In the case studies of the first half of the course, the United States appears very little. We want officers to cultivate a penchant for dispassionate strategic thinking before they move into the second half of the course, where the United States looms large and where passions might otherwise swell high.

Admirals who have not taken our course, and other distant onlookers unfamiliar with our approach, often ask why we start all the way back in ancient Greece, with a war fought with rowboats and spears and chronicled in a very long work of historical inquiry in its intellectual infancy. Preoccupied with the application of technology in military operations, they overlook the fact that our course is about strategy on a grand scale. They also do not see that Thucydides's account of the war between Athens and Sparta still stands as the greatest work of strategic history ever written.[23]

Thucydides provides historical backing to Clausewitz's conceptual picture of war as a realm of fog and friction, chance and uncertainty, and dynamic interaction between adversaries; and also to his trinity concept, with its ever changing mixture of rational political purpose, military creativity (and blunders), and human passions that can deflect both rational control and creative use of violence. Thucydides has the further virtue for strategy and policy of giving us much food for thought as we digest our 11 course themes. Looking at both Athenian and Spartan policies and strategies, officers can begin the hard task of thinking about what exactly makes for a policy-strategy match and what forms a mismatch might take. Thucydides takes them deep into the decisions for war on both sides and highlights causes of war that political scientists still debate in our day. In the debates at the time, both the Athenian leader Pericles and the Spartan leader Archidamus did net assessments. Students can see that Archidamus's assessment was superior because it enabled him to anticipate the nature of the war as it was to unfold, but they also note that his cautionary perspicacity did not keep Spartan hawks from driving their city-state into war.

As they immerse themselves in the Greek fighting, and move from SLC's process themes 1-3 to 4-7, today's officers get a classic introduction to a fundamentally asymmetric war: the Athenian whale, with its superior naval power, had to figure out a way of strategically overcoming or outmaneuvering the Spartan elephant, with its superior ground force. At the theater level, the Peloponnesian War provides several examples of enduring value in the 21st century about how to master asymmetry by operating on favorable terms in the enemy's preferred domain. But such operational successes did not translate directly into durable strategic success in the war between Athens and Sparta. Finding it difficult to come up with an overall strategy that could deliver their desired political outcomes, both sides had to engage in reassessments that involved opening or contesting new theaters in the ongoing war. One such decision to open a new theater — the Athenian expedition to Sicily in 415 BCE — proved to be a major turning point in the war. Rear Admiral Wisecup, the current president of the Naval War College and a student in Strategy and Policy more than a decade ago, still has vividly in his mind the failure of Athens in Sicily as an object lesson for the United States in the 21st century.

Both before and after the Sicilian fiasco, Athens had opportunities to terminate the war with negotiated compromises. In analyzing these opportunities, students encounter three enduring issues of war-termination policy and strategy: how far to go militarily; what to demand politically; and how to secure in a postwar era whatever favorable terms a peace settlement has established. In this case, how far to go militarily meant, for Athens, how often it should risk its fleet in distant waters. In the event, Athens fought to the bitter end, losing its fleet and thence everything of value to its city-state. But moving on from Thucydides to Plutarch's account of Lysander, the talented Spartan commander of the new fleet that enabled victory over Athens by an elephant that had become a whale as well, students see how Sparta then lost the peace and ended up in new wars that eventually brought it to ruin.

The four SLC environment themes help officers deepen their strategic understanding of this Greek tragedy. In the international dimension of strategy, Athens started with a large coalition that it assessed as superior to the Spartan coalition. But in the war, Sparta did a better job of both maintaining coalition cohesion and adding third parties to its side. Students here see the beginning of a pattern that points toward a key factor of success in high-stakes, multitheater wars: the side that puts together the most extensive and cohesive coalition has a much higher probability of ultimate victory than the other side. In the material dimension of strategy, the Athenians started with a vastly superior economic system—a capitalist system with globalized markets throughout much of the Greek world—but it ended up losing the war because of economic effects; they were starved into submission by a naval blockade and ground siege of their city. The development and maintenance of economic superiority is another key success factor in the more recent big wars studied in the SLC.

The institutional dimension of strategy is emphasized by Thucydides. After Pericles died early in the war, the political competition for leadership in the Athenian democratic system got so ferocious that it led to self-defeating actions in the realm of strategy and dysfunctional command relationships in the realm of operations. American officers are quick to debate how their political system may be similar to, and different from, the Athenian political system in terms of strategic behavior. In the cultural dimension of strategy, we can see how, for all the common culture shared by the ancient Greeks, they still had trouble not only in cooperating with each other politically — not least because of ideological divisions and other passions — but also in anticipating how one city-state would react strategically to what another city-state did. In the social dimension of strategy, future strategic leaders can mull over how the capacity for innovation that enabled Lysander to lead Sparta to victory in the Peloponnesian War turned out to have a very corrosive effect on the long-term viability of traditional Spartan society.

After our ancient Greek rite of passage, students and faculty in the SLC leap forward 2 millennia (much to the chagrin of historians) and then hop around the historical landscape of the modern world. There is a method to our seemingly erratic madness: we select cases with an eye to giving students broad and balanced exposure to different types of war. Those types fall into three boxes of S&P wars:

- Box #1: Big wars (or global wars) — high-stakes (usually unlimited political aims), multitheater, protracted coalition wars;
- Box #2: Regional wars — lower stakes (usually limited political objectives), single theater (or perhaps two contiguous theaters), rarely so protracted, often with one side isolated but with third parties perhaps on the verge of intervention;
- Box #3: Insurgencies — nonstate actors seeking to overthrow or secede from an established political system and form a new state or a more amorphous political entity.

Rather than concentrate on types of warfare in terms of the forces engaged — regular, irregular, and hybrid — or in terms of tactics employed — conventional, guerrilla, and terrorist — we envision types of wars in more political terms. The political essence of the big wars is about who will dominate the distribution of power and define the norms of the international system in which the major players seek to advance their interests. The political essence of the regional wars is typically about who will control crucial pieces of territory in one region of the world. The political essence of insurgencies is about the nature or extent of a national or imperial political system. The different types of warfare around which so much DoD debate centers these days can be present in any of these boxes. By the same token, nonstate as well as state actors can be involved in big wars and regional wars as well as in insurgencies, as we see today with armed jihadist groups.

In the 3 months of the SLC, officers get intellectual exposure to seven big wars and to about twice that number of both regional wars and insurgencies. In the 1990s, when many other JPME departments were reducing their attention to insurgencies, the Strategy and Policy department increased the number of its case studies in that box of wars. Currently, when there is limited attention to possible big wars beyond the Global War on Terror, the department intends to keep up its countervailing ways and ensure that future strategists who are studying at Newport will be as well-prepared intellectually for the big-war box as for the other two boxes of wars.

Boxes imply separation. But as case studies reveal, the S&P boxes of wars may overflow or overlap. A war may start in one box and end up in another box. There may be wars within wars or, as in the Vietnam case from 1965 to 1972, all three boxes may be in play. In the first decade of the 21st century, the United States has again had to try to handle all three boxes, making transitions from regional wars to insurgencies, all in the context of an overarching big war. Better to prepare officers for such messy reality, all SLC historical cases feature more than one type of war. In some cases the types are simultaneous; in some cases they are sequential, with a transitional period between the different wars. A recurring pattern in the cases has a major power handling one box well and then stumbling badly in the next box, as happened with the United States after 9/11.

As SLC cases pile up in the three boxes, officers should be able to see how each new case shares some common features with the previous cases in its box but also presents some distinctive new features as well. Thus, there are both old patterns and new puzzles to consider every week. For each case in a box, officers should be able to explain why one side won and the other side lost. Those explanations will have generic factors in common with the other cases in the box and idiosyncratic factors not in evidence in the previous cases.

Periodically, at the end of a seminar meeting, I ask students if they see some generic success factors for each box as cases have accumulated in it. My discussion of the Peloponnesian War in this chapter highlighted three such success factors for that big war: coalition cohesion, development and integration of different instruments of power, and economic superiority. One might add an Athenian propensity to engage in self-defeating action, with the Sicilian expedition marking the road to strategic overextension. It turns out that these factors show up again and again in the big wars of the SLC. I shall leave it to students and readers (and my own book in progress) to figure out the different sets of success factors for the other two boxes of wars.

The point in undertaking this sort of exercise is not to leave officers with yet another foolproof checklist to gather dust on top of the others that they have collected in their military careers. Factors of success do not guarantee victory in war. They are probabilistic, not deterministic. Cultivating those factors improve the odds of success, but never to 100 percent. Idiosyncratic factors, as well as chance and contingency, always matter in the real world of war. But if in a future war, strategists educated at the Naval War College encounter arguments for courses of action that cut against the success factors for the box in which that war falls, they should be ready, willing, and able to mount powerful counterarguments.

MORE PATTERNS

Factors of success in the different S&P boxes of wars are not the only types of patterns that students and their faculty moderators can discern as they make their way through the term. Patterns can be found in the course themes as they are applied to the weekly case studies. For example, from the perspective of theme 1 ("The interrelationship of policy, strategy, and operations"), participants in the course see eight basic types of mismatch between policy and strategy popping up again and again. Repeated exposure to theme 2 ("The decision for war") reveals that, typically, political and military leaders in all times and places proceed into violent interaction with the enemy without developing, agreeing upon, and articulating a realistic theory of victory—by which I mean, in this context, a set of assumption about how the military operations that they contemplate will translate into the political outcomes that they envision. Focusing more narrowly on the U.S. experience as a global power in the past century in terms of themes 5-7 ("interaction, adaptation, and reassessment"; "war termination"; "winning the peace and preparing for war") highlights another recurring pattern: the considerable difficulty that, with rare exceptions, American strategic leaders have had in handling phase transitions both within a war and across the cycle running from war to peace to the next war. Much of this difficulty stems from some combination of a lack of intellectual forethought and an abundance of institutional friction.

As one turns from the process themes to the environment themes, patterns with recurrent motifs become apparent there as well. For example, in the coalition component of theme 8 ("The international dimension of strategy"), one repeatedly sees the salience of certain basic military and nonmilitary elements working for or against coalition cohesion. Similarly, with regard to the key part of theme 10 ("The institutional dimension of strategy") that deals with civil-military relations, one comes to comprehend the common sources of friction that have so frequently made for strategically dysfunctional interaction between political leaders and military leaders.

Patterns emerging from the course themes with such regularities and recurrences are not only the easiest to grasp, but also have the greatest probability of persisting in the future. Alas, they do not, by and large, point in any straightforward way toward prescriptive lessons on how to achieve strategic success; rather, for the most part they put up cautionary signposts about strategically self-defeating behavior to avoid. Other more dynamic patterns have a degree of variation that makes them chancier to project forward in time. But they can provide macroscopic situational awareness about predicaments in which commanders and high-level staffers might find themselves in the future.

Such dynamic patterns become conspicuous in theme 4 ("The instruments of national power") as students progress through the course. Many ground and air commanders have aspired to quick decisive victories ("QDVs" in Newport's Strategy and Policy parlance). Relatively few have achieved them. Historical case studies looking at the rise and fall of Napoleon and then of Imperial Germany help officers see that the reasons for success and failure in such an endeavor go beyond the operational skill of commanders in maneuvering their forces. Policymakers, whose wisdom may wax and wane, must carefully calibrate their political objectives. Technological and geostrategic circumstances, which vary from one war to another, must allow for some major qualitative or asymmetric advantage in military instruments. The international environment, which can be like a kaleidoscope, has to permit the diplomatic isolation of the adversary. The enemy may or may not be like a Bologna flask (to use Clausewitz's metaphor); some types of enemy political systems are more likely than others to implode or give in politically after they have suffered one or two big operational blows early in a war.[24] In a future regional war, strategic leaders in quest of QDVs ought to be aware of these critical variables.

In a protracted war involving insurgency (which, as we were reminded in Operation ENDURING FREEDOM (OEF) and Operation IRAQI FREEDOM (OIF), may emerge after the quick collapse of a defeated regime), the situational awareness of commanders and their subordinates must be reoriented toward the people caught between a failing, emerging, or established government on the one side and the insurgent movement on the other side. SLC cases bring to the surface the many different types of circumstantial factors or combatant actions that have swayed the alignment of the people one way or the other. The relative importance of different factors has varied over time and across different environments. So has the strategic mix of insurgent actions—guerrilla, conventional, terrorist, and information operations—and political purposes. SLC cases also reveal the types of features of an environment that play to the advantage of the insurgents or of the government and, not least, the crucial importance of the relative value of the object for the two sides.[25] Strategic leaders need to be prepared to find ways not only to assess (SLC theme

2) and attack the enemy's strategy (Sun Tzu applied to counterinsurgency), but also to mitigate the unfavorable environmental features and to communicate about value to audiences who may question what is at stake in the conflict. This complex of considerations is challenging for students to assimilate intellectually, but if they can meet the challenge in the classroom, they should be well-situated in the battlespace to adapt to the dynamics of insurgencies that they will encounter as senior commanders and staff officers.

Macroscopic situational awareness presents even more of a challenge for maritime commanders. Whereas the United States has had much recent experience in wielding ground and air instruments against insurgencies and in regional wars, it has not fought a major fleet-on-fleet engagement since 1944 and has not executed a major forcible-entry amphibious operation since 1950. Whether senior U.S. Navy leaders have appreciated the fact or not, lack of experience in major warfighting missions against competent adversaries puts an especially great premium on the need for intellectual preparation on the part of future maritime commanders and staff officers.

In both the ILC and the SLC, future fleet commanders can see that fleet-on-fleet engagements have, historically, clustered in intervals of several decades (1759-1805, 1894-1944) separated by even longer intervals in which the dominant naval power has gone unchallenged on the high seas. Overarching this patchy pattern are two very long-term trends (which I point out in a lecture on maritime strategies in retrospect and prospect): warfare at sea has become ever more lethal, and fleets have become increasingly lumpy, in the sense that more and more combat power is invested in fewer and fewer capital ships. Examination at the end of the course of possible maritime futures suggests that a new era of high-seas, high-stakes naval conflict may be looming in the 21st century and that the proliferation of missiles may make it possible for hapless naval commanders to lose a very expensive fleet in a very short time.

Also, in both the ILC and the SLC, future amphibious commanders learn from historical cases that there has been a long pattern of oscillation in the offense-defense balance between forces coming from the sea to the land and forces seeking to deny access to such maritime power projection. This pattern has arisen primarily (though not exclusively) from technological change. In the 21st century, the impact of the proliferation of missiles, along with advances in mines, submarines, and means of surveillance, seems quite likely to give a major edge to integrated coastal-defense systems of competent military establishments. This pattern projection should serve as a precautionary signal to our would-be maritime commanders to enhance their awareness of technological tides and ponder the ways in which they might handle the risks that the tides may portend.

Seeking in the SLC to balance situational awareness about the instruments of military power and situational awareness about non-military sources of national capability, the Strategy and Policy faculty has been considering how best to incorporate more deeply into the course a major problem in the institutional dimension of strategy (theme 10) and a disquieting possibility in the material dimension of strategy (theme 9). With regard to the former, a pattern of interagency incoherence in the planning and execution of strategy has given a pronounced American institutional twist to Clausewitz's notion of the pervasive role of friction in war. On the positive side, we do offer to students some examples of fruitful interagency effort, mostly from the 1940s and early 1950s. On the negative side, we ask students to read Robert Komer's trenchant analysis of interagency shortcomings in the Vietnam War.[26] Yet, despite his lamentations and lessons learned, the problem seems only to have gotten worse for much of the American involvement in Iraq and Afghanistan. Here, too, military students and civilian officials can emerge from the course with enhanced situational awareness of, but not with tidy solutions to, the problem. After studying in an academic setting the few ups and many downs of this institutional pattern, they will have to figure out ways to make their relationships work better in their future assignments.

The United States has usually been able to compensate for deficiencies in the institutional dimension of strategy with superiority in the material dimension of strategy. Of course, clever adversaries will seek to sidestep or blunt that superiority, as the Vietnamese communists did, and as Usama bin Laden hopes to do with an al Qaeda plan to bleed the United States to bankruptcy.[27] But if the United States finds its material superiority greatly diminished in the future, it will be largely because of deeper-seated patterns that the Strategy and Policy course is now giving closer attention. Taking as a point of departure Paul Kennedy's celebrated book on *The Rise and Fall of the Great Powers* (1987), we expose students in two different case studies to the long-term relationship between British economic and strategic performance. The point is not to draw a simple historical analogy between Britain and the United States, but rather to introduce economic concepts — such as total factor productivity and growth accounting — and economic conditions — financial crises, fiscal crunches, and dependence on critical overseas economic inputs — that illuminate how some types of material constraints that plagued British strategy in the past may come increasingly to affect American strategy in the future. The macroscopic situational awareness on offer here to future strategic leaders is that they may have to operate not only within manpower constraints (as they are quite used to doing), but also within broader material constraints (to which they are much less accustomed). Such a constrained future will make the ability to come up with concepts that are both creative and realistic an even more important trait of strategic leadership than it already is.

SOME PUZZLES

Anyone who thinks in terms of patterns should also think in terms of puzzles. Here I have in mind radical discontinuities, major breaks in patterns, which pose difficult problems for strategists to handle. Twenty-first-century puzzles lurk in the three S&P boxes of wars. A common thread connects puzzles across the boxes: the spread of weapons of mass destruction (WMD) or disruption. When I joined the Strategy and Policy faculty in the 1980s, the Cold War was not yet over, and nuclear weapons were part of the course taught then. In the 1990s, nuclear weapons receded well into the background of the curriculum. They have now reappeared, most conspicuously in a new case study on the proliferation of nuclear weapons. Its primary focus is on how to employ DIME in an effort to prevent the spread of nuclear weapons, but looming over the horizon is the issue of wars between belligerents that already possess them.

Proliferation optimists hypothesize that nuclear weapons have made high-stakes, multitheater, coalition wars a relic of the past. The strategies of an unlimited big war like World War II in Europe, with large armies marching on capitals and with massive strategic bombing of cities, would not meet the test of rationality if the object of such attacks possesses nuclear weapons and the means to deliver them over long distances. But multitheater maritime-heavy wars between nuclear powers are not so unthinkable. The outcome of such wars in the past between non-nuclear powers has often turned on the key strategic issue of when and where to open new theaters, and one can imagine a big maritime war in the future extending around the eastern and southern rimlands of Eurasia. One of the noteworthy contributions of the SLC to thinking ahead about such a war is a framework (which can be adapted to other types of war as well) for assessing whether or not one can be operationally effective in, and gain a positive strategic payoff from, a new theater. The puzzle to solve here is how in effect to get the opposing side to defeat itself through strategic overextension.

Some proliferation optimists have questioned whether even a more limited regional war between states with nuclear weapons is likely in the future. The Kargil conflict unleashed by Pakistan in 1999 against India in Kashmir would suggest that the answer is yes.[28] Indeed, students finishing their SLC excursion around the regional-war box should proceed on the supposition that most future regional wars involving the United States on one side will have a nuclear-armed state on the other side. The types of strategies that have delivered quick decisive victories in past regional wars might well be deemed too risky in such future regional wars. In that case, the puzzle to solve will be how to exert sufficient war-termination leverage to bring the fighting to an end on favorable terms that have some prospect of being durable. The SLC has another general framework to contribute here, this one about war-termination strategic options, which future strategic leaders should be able to adapt to the specific circumstances of a regional war against a nuclear state in the years to come.[29]

The shock of the 9/11 attacks and intelligence about a meeting between Pakistani nuclear scientists and al Qaeda leaders concentrated the minds of American policymakers and analysts on a dangerous discontinuity in the insurgency box. If AQAM are regarded as a global insurgency, the pattern of insurgencies has taken a quantum leap in political ambition, geographical reach, and destructive power. The puzzle to solve here takes us back to the mega-concepts of rationality and interaction. The odds of a WMD detonation brought about by AQAM in an American city over the next decade or so are quite uncertain. Even if the probability is assumed to be low, the consequences of such an eventuality would be horrific. In such a situation, there is a fundamental riddle of rationality: it is hard to know what costs and risks are worth incurring to prevent a low-probability, high-consequence event. Because AQAM leadership is hard to find and finish off, the interaction dynamic is fraught with difficult trade-offs for American strategic leaders. Are the short-term operational prospects for unmanned aerial vehicle (UAV) strikes positive enough to be worth the risk of negative political effects within Pakistan, effects that might play out in an implosion of the current regime in Islamabad? Alternatively, is the value of the object in Afghanistan, Pakistan, and other theaters worth the cost in magnitude and duration of counterinsurgency campaigns designed to deny sanctuary to AQAM? How will the value and the costs be assessed and reassessed over time within the American political system? Is there any strategic concept other than the current operational mix of UAV strikes and counterinsurgency campaigns that might work better in terms of rationality and interaction? It is such advanced strategy and policy questions that we all would like the course to educate military officers and civilian officials to answer insightfully.

CAN GREAT EXPECTATIONS BE FULFILLED?

Enough has been said in this chapter to indicate how far-reaching and challenging the SLC in Newport has become. Patterns and puzzles, factors and frameworks, course themes and theoretical concepts, historical cases and generic boxes of wars make for a complex intellectual infrastructure, though the degree of abstraction does not approach the murkiness found in most upper-level social-science courses in the civilian academic world. By no means does every seminar probe with equal depth into these underpinnings of the SLC, and by no means do all students fathom them.

Graded exercises provide a measure of who among the students can grasp and then use which of the tools laid out in the course. The distribution of grades takes the familiar bell-shaped configuration. The tail on the left of the curve, the bottom quarter of the grades, is perhaps best passed over in silence. The big bulge in the middle consists of students who have respectable skills of communication and serviceable competence in the selection and execution of appropriate strategic concepts. Moving on from Newport, they should at the very least take with them not only a much enhanced situational awareness, but also a good sense of what strategically self-defeating behavior is and how to avoid it. That is not faint praise. So long as the United States can avoid any further outbursts of self-defeating behavior, there is no compelling reason for the strategic pessimism that enshrouded the beginning of this chapter.

It is the tail on the right of the bell-shaped grading curve, the top quarter of the students in the course, which provides the best test of the proposition advanced by Krepinevich and Watts about the limits of what we can expect from an education in strategy. The Strategy and Policy faculty at the Naval War College have greater expectations than they do. We think that we can help organize the minds of students in ways that will make them much more creative strategic problem-solvers than they were before they came to Newport. The students who separate themselves from the bulge in the middle of the curve can create new ideas from old books, above all from Clausewitz's method of critical analysis of strategic alternatives and Sun Tzu's maxim about attacking the enemy's strategy and alliances. In future positions of great strategic responsibility, our best students should be able to use their minds not just to avoid self-defeating behavior, but to accelerate progress on the path to victory in war.

As Congressman Skelton has pointed out, whether JPME courses are creating strategic thinkers is not the only important question. Of equal importance is a further question: are the military services identifying such thinkers and using them in optimal ways? To a faculty member in a course that sorts out the best strategic minds, it is disheartening that the services seem to pay so little attention to the indicators of excellence that we can provide. To be sure, intellect and formal education are certainly not sufficient for senior officers to be first-rate strategic leaders. Also necessary are temperament, experience, and a personal effort of self-education after the formal education comes to an end. Three months in Strategy and Policy classrooms can only be a point of departure on a long strategic journey.

ENDNOTES - CHAPTER 5

1. Barry D. Watts, "US Combat Training, Operational Art, and Strategic Competence: Problems and Opportunities," Washington, DC: Center for Strategic and Budgetary Assessments, 2008; Andrew F. Krepinevich and Barry D. Watts, "Lost at the NSC," *The National Interest*, January 6, 2009, available from *www.nationalinterest.org/Article.aspx?id=20498*; Andrew F. Krepinevich, "The Project on National Security Reform: Challenges and Requirements," testimony before the U.S. House of Representatives, House Committee on Armed Services, Subcommittee on Oversight and Investigation, March 19, 2009.

2. Richard K. Betts, "Is Strategy an Illusion?" *International Security*, Vol. 25, No. 2, Fall 2000, pp. 5-50. See also Richard K. Betts, "The Trouble with Strategy: Bridging Policy and Operations," *Joint Force Quarterly*, No. 29, Autumn/Winter 2001-02, pp. 23-30.

3. Bradford A. Lee, "Winning the War but Losing the Peace? The United States and the Strategic Issues of War Termination," in Bradford A. Lee and Karl F. Walling, eds., *Strategic Logic and Political Rationality*, London, UK: Frank Cass, 2003, pp. 249-273.

4. The importance of strategic assumptions is at the heart of a long manuscript that I am currently preparing with the immodest title of "On Winning Wars." The work lays out and examines theories of victory. One meaning of "theory" in this context is the assumptions that strategists make about how the military operations being planned will translate into the political objectives for which they are fighting a war.

5. Rear Admiral Jacob Shuford, USN (Ret.), "Re-Education for the 21st-Century Warrior," *U.S. Naval Institute Proceedings*, Vol. 135, No. 4, April 2009, pp. 14-19.

6. In the course critique submitted at the end of the academic term, officers give the highest marks, usually by a wide margin, to the 2 weeks in which they are asked to read the oldest and most difficult books: Clausewitz, Sun Tzu, Thucydides, and Plutarch. They give much lower marks to the 4 weeks that cover ongoing wars and emerging problems of the future—weeks in which they read analyses and accounts unlikely to meet the test of time.

7. Vice Admiral Stansfield Turner, "Convocation Address," 1972, reprinted in *Naval War College Review*, Vol. 51, No. 1, Winter 1998.

8. Lee, "Winning the War but Losing the Peace?" pp. 262-263.

9. Philip A. Crowl, "The Strategist's Short Catechism: Six Questions without Answers," *The Harmon Memorial Lectures in Military History*, No. 20, Colorado Springs, CO: U.S. Air Force Academy, October 1977, pp. 1-14.

10. Carl von Clausewitz, *On War*, Michael Howard and Peter Paret, eds. and trans., Princeton, NJ: Princeton University Press, paperback Ed., 1989; and Sun Tzu, *The Art of War*, Samuel B. Griffith, trans., Oxford, UK: Oxford University Press, paperback Ed., 1980. We do assign a survey of contemporary strategic theory conducted expertly by Lawrence Freedman, *The Transformation of Strategic Affairs*, Adelphi Paper 379, London, UK: International Institute for Strategic Studies, 2006, in order to have the students taste its flavors of the day. The taste test is not favorable for the moderns against the ancients.

11. Both the Griffith translation of Sun Tzu and the Paret/Howard translation of Clausewitz are more literary than literal. Accordingly, those translations are relatively accessible to students, but are open to serious question in their rendering of parts of the two texts.

12. See Book 2, Chapter 5, of Clausewitz, *On War*.

13. The quotations are from Clausewitz, *On War*, pp. 159, 167.

14. *Ibid.*, p. 605.

15. The concept of effects appears much more frequently in *On War* than does the concept of center of gravity, and in one passage (pp. 485-486) the former concept is embedded in the latter one. Without linking Clausewitz's traditional notion of effects to recent U.S. notions of effects-based operations, which Clausewitz surely would have regarded as an abomination, it is often productive to encourage students to explore the range and intensity of strategic effects that different courses of action might generate.

16. On the important strategic issue of deploying forces to a secondary theater, sometimes far from the central theater, we contrast Clausewitz and Sir Julian S. Corbett, whose book *Some Principles of Maritime Strategy*, London, UK: Longmans, 1911, I shall comment on later.

17. Michael I. Handel, *Masters of War: Classical Strategic Thought*, 3rd Ed., London, UK: Frank Cass, 2001, pp. 181, 183.

18. Our late colleague Michael Handel in his *Masters of War* (excerpts from which are assigned in the first week of the course) emphasized common ground between Clausewitz and Sun Tzu. Most Strategy and Policy faculty members are more inclined to highlight the differences.

19. X [George Kennan], "The Sources of Soviet Conduct," *Foreign Affairs*, Vol. 25, No. 4, July 1947, pp. 566-582.

20. Sun Tzu, *The Art of War*, pp. 77-78 (menu of options), 86-90 (keys to victory), 91-92 (*cheng* and *ch'i*).

21. Last year the calendar allowed us to fit in an 11th case study, on the end of the Cold War. Next year we hope to add a different case study on South Asia, with special attention to the vicissitudes of the troubled relationship between Pakistan and India.

22. Admiral Shuford, "Re-education," p. 16, quotes the old saw that "[e]ducation is what's left when you have forgotten all the facts."

23. A recent edition enables students to make their way through Thucydides much more easily than before. R.B. Strassler, ed. *The Landmark Thucydides*, New York: The Free Press, 1996. One might argue that the modern historical work most closely resembling our strategy and policy approach is also about the war between Athens and Sparta: Donald Kagan's four-volume history of the Peloponnesian War, published by Cornell University Press between 1969 and 1987.

24. The Bologna flask metaphor is in Clausewitz, *On War*, p. 572. The flask is so brittle that it easily shatters when scratched.

25. On the value of the object, see *ibid.*, p. 92.

26. Robert Komer, *Bureaucracy Does Its Thing: Institutional Constraints on U.S.-GVN Performance in Vietnam*, Santa Monica, CA: RAND Corporation, 1972.

27. Translated transcript of a Usama bin Laden video, posted on al-Jazeera's website, November 1, 2004.

28. See the discussion of the Kargil conflict in S. Paul Kapur, "Ten Years of Instability in a Nuclear South Asia," *International Security*, Vol. 33, No. 2, Fall 2008, pp. 71-74.

29. Some of the contours of this framework can be deduced from Lee, "Winning the War but Losing the Peace?"

CHAPTER 6

A VISION OF DEVELOPING THE NATIONAL SECURITY STRATEGIST FROM THE NATIONAL WAR COLLEGE

Cynthia A. Watson

The National War College (NWC) has been educating future civilian and military leaders of the national security community since opening to its first class in September 1946. Its mission "is to educate future leaders of the Armed Forces, State Department, and other civilian agencies for high-level policy, command, and staff responsibilities."[1] Ambassador George F. Kennan held the first position as International Affairs Advisor in the College in 1946 where he penned the "Mr. X" article in *Foreign Affairs* that became the basis of the containment doctrine against the Soviet Union. Kennan is one of a long line of distinguished statesmen that have taught and studied at the NWC.

Over those 7 decades, the student body and curriculum have evolved to include a range of agencies and topics, respectively, many of which were not of interest to the strategists of the Cold War (1947-92). As with other senior service colleges (SSCs), NWC students serve at Fort Lesley J. McNair in southwest Washington, DC, for 10 months, receiving upon completion a master's degree in National Security Strategy that is accredited by the Middle States Association of Colleges and Universities. Military officers also come to the College under the auspices of the J-7, Operational Plans and Joint Force Development, of the Joint Staff. Both Middle States and the J-7, with its Officer Professional Military Education Policy

(OPMEP), which is periodically updated, reevaluate the College's delivery of education to its students relative to the goals the College's mission sets forth. Periodic reviews are one, but not the sole, method of the National War College assessing its ability and efficacy in educating national security strategists. This chapter discusses the approach that NWC takes to teaching strategy.[2]

FACULTY RECRUITMENT AND DEVELOPMENT

The faculty has three components: uniformed military, agency faculty, and civilians known as "Title X" faculty for the portion of U.S. Code under which they are hired. Uniformed faculty constitute roughly a third of the faculty, split equally across the two administrative departments. From the beginning in the 1940s, the College has had an equal split of Army, Air Force, and Sea Service faculty (with Marines and Coast Guard included in the latter). These faculty members are generally recruited at the 25-year mark in their careers after vast operational experience. If possible, the College recruits from the shrinking pool of faculty holding doctorates, but today's officers often only hold a masters' degree. Faculty members come for 2- to 3-year periods from their Services. As true with any of the faculty of the College, a high premium is placed on those who have prior teaching experience but new faculty receive considerable development, as shown below.

Inter-Agency faculty also spend roughly 2 or 3 years at Fort McNair. The State Department has been a stakeholder in the College since the 1940s when Ambassador George F. Kennan was the first International Affairs Advisor, and State has the largest number of agency faculty along with the single largest number of students from any single agency. The agency faculty, as they are collectively known, are similarly senior in their fields, bringing equally impressive operational credentials. Most of the agencies that send students to the College also send a faculty member under a memorandum of agreement to represent the agency's role in the national security strategy making process. In recent years, members of the faculty have come from the Department of State, the Office of Secretary of Defense, the intelligence agencies, the Defense Intelligence Agency, the National Security Agency, the Federal Bureau of Investigation, and the Department of Homeland Security. The desire is to recruit faculty with doctorates, but the faculty are secunded by their agencies, which means that the doctorate is not always available.

Title X faculty are on renewable contracts and are recruited from across the world through a competitive process. The overwhelming majority of faculty members have doctorates in a field related to national security strategy, while two are lawyers at present. An equally important desire is that faculty hold some practical experience related to national security, such as work on Capitol Hill, in uniform at a combatant command, or in a high level decisionmaking job. While there are a couple of faculty members who are pure academics, the fear has always been that someone without the expertise in government would be at a distinct disadvantage in seminar discussions with NWC students who put such emphasis on practical experience as they discuss course material. Title X faculty teach but are also expected to conduct research to maintain a currency and freshness in teaching as well as a foot in the national debate on security issues.

All faculty undergo a new faculty orientation and faculty qualification program. The week-long orientation introduces the basics of the College. The Qualification Program allows each new faculty member to be an observer (backseater) in at least two core courses before doing solo teaching. The backseater experience allows time to see various teaching styles, as they are each assigned to a mentor (whom they generally observe regularly) but also encouraged to attend other seminars to see how other faculty members' styles work in a classroom. Additionally, backseating faculty conduct at least two seminars per core course to allow the primary faculty member and mentor to see the quality of faculty preparation, techniques, and such. The mentor certifies to the College leadership that a new faculty member is prepared to lead her or his seminar before such an assignment is made. All faculty also attend workshops to help with substantive discussions on seminars as well as to share pedagogical techniques.These workshops run weekly in the fortnight before and throughout a core course. Finally, the Director of Development also holds periodic brown bag lunches on pedagogy and substantive topics of interest to all faculty, with some of the topics, such as evaluation and assessing writing, aimed primarily but not exclusively at new faculty members.

DEVELOPING, ADMINISTERING, LEADING SEMINARS, AND EVALUATING CURRICULA AT THE NWC

It would be a far simpler world if there were an easily identifiable body of thought on what exactly needs be taught in national security strategy. National security strategy is the orchestration of all of the instruments of national power in pursuit of achieving national interests, based on an understanding of how that strategy can be advanced within the U.S. national political, social, and economic context along with the milieu of the international situation. The overall objective of the College thus requires grasping the national interests being served by any particular strategy, identifying the instruments of statecraft which are both applicable and inappropriate in scope, determining how the various portions of the U.S. body politic react to goals and possible uses of the instruments, and recognizing that the United States (or any other strategic actor anywhere in the world) is part of a broader context of concerns over which it does not have complete control. This relatively wide-ranging agenda forms the basis of what new strategic thinkers must confront in the tight 10 months they study at the College.

Students arrive at the College through one of three paths. Sixty percent of the average class of 223 are officers equally divided among the Sea Services (Navy, Marine Corps, and Coast Guard, even though the latter is formally part of the Department of Homeland Security rather than the Department of Defense [DoD]), the Air Force, and the Army. These students are predominantly at the lieutenant colonel grade for Marines, Air Force, and Army (the equivalent rank for Navy students is commander) but with a number of colonels/captains, who come through nomination by their Services as part of their professional career development and as a stipulation of the Goldwater-Nichols Military Reform Act of 1986, which requires professional military education (PME) at Level II, which the NWC has been offering for 2 decades. These students have excelled as operators in their specialties, whether as a Marine infantry officer, Army helicopter pilot, Navy surface warfare officer, or Air Force intelligence officer. The Services have given the students the opportunity to provide input into the slating for schools but the final determination resides with the personnel branch of the individual Services. Students, averaging 19 years' experience in uniform, come by designation of the Service.

Civilians form 40 percent of the incoming class. They may come as a result of their agency's selection or they may be self-nominated. One change in the first years of this decade was the inclusion of larger numbers of civilians from outside of the State Department, which is an original stakeholder in the NWC.[3]

The third path to the College is for the 32 international officers who receive invitations from the Chairman of the Joint Chiefs of Staff to attend as representatives of their nations. Unlike the U.S. students, these may come without undergraduate degrees, in which case they participate in classwork but are not eligible for the Master's Degree in National Security Studies.

The students thus have been through a screening process for admission that differs from the traditional preparation criteria. The selection for the majority comes from their Services' views of the component parts that go into the creation of a strategist, including leadership skills and prior assignments. The effect that this selection criterion has on teaching strategy at the NWC is that the College's goals for the 10-month academic program are broader than simply the seminar learning experience. The NWC experience promotes the development of national security strategists through (1) seminar-driven learning, reading, and discussions, (2) greater knowledge of capabilities not only across and within the Services but across the federal government and with potential partner nations, (3) expansion of leadership capabilities through participation in multifaceted case studies, and (4) lectures and discussions with high level civilian and military officials of the United States and foreign countries.

The point to discussing access is to underscore that students have already achieved a level of professional success before arriving, hence the College curriculum is quite different from that of other post-undergraduate schools. The NWC (and other SSCs[4]) is more aptly described as a professional school than a traditional graduate school. One aspect of this professional school is the requirement to incorporate as much current information, be it through readings, speakers, or case studies, as possible. Otherwise, students see little applicability to their imminent return to their jobs as national security practitioners.

The curriculum undergoes annual and course-by-course evaluation by students and faculty. At the end of each core course, students offer unfettered written quantitative and qualitative responses to the questions on the course methodology, content, readings, instruction, and other dimensions. Additionally, students offer their comments to a single seminar member who represents the seminar at a discussion session, known as a hot wash, in which each seminar is represented and where points of discussion are teased out. Finally, faculty members participate in an evaluation of the individual core course with the course director. Course directors then use the student and faculty feedback to revise and update their courses for the following year. The Director of Assessment and Institutional Research prepares a written analysis of the course feedback for the Commandant and College leadership.

Additionally, the College continues to reevaluate its curriculum through several other methods. A Curriculum Coordinating Committee, consisting of the directors for all seven of the core curriculum courses, Associate Dean of Faculty and Academic Programs, Dean of Faculty and Academic Programs, and the two department chairpersons, along with the Director of Assessment and Institutional Research who sits in *ex officio* status, meets regularly through the calendar year to discuss the course content including specific topics, readings, and speakers to avoid duplication as well as to prevent gaps on crucial topics.

NWC students also take hour-long oral evaluations twice during their program. These are opportunities for the student to demonstrate her/his learning by engaging with two faculty members. At the same time, it also allows faculty to understand far better how well strategists react to questions posed to them much as a commander might rely on a staff officer. The evaluation tests not only the student's ability to respond to discrete questions but also allows the evaluators to monitor the strategic logic the student is using. These are some what stressful, much as a high-level strategy position would require for a real-time position. In more than 15 years of conducting these evaluations, the College has noted some deficits in its curriculum and adjusted accordingly to make certain it is meeting the nation's needs as a school for strategists. One of the startling voids that oral evaluations revealed in the spring of 2003, for example, was a fundamental lack of student understanding of the importance and complexity of nation-building in strategy. This revelation appeared slightly ahead of the dawning of this deficit in the strategy of the nation in the aftermath of invading Iraq in March of that year.

The College also submits questionnaires to graduates at regular junctures after they leave and it also surveys the senior stakeholders to see how the graduates perform in their post-graduate positions. This feedback is part of the material the leadership assesses in evaluating the need to revise the curriculum.

A formal, independently constituted curriculum review occurs every 3 to 5 years under a committee including all the component faculty of the College: uniformed officers, agency faculty on loan, and Title X civilian academics with professorial ranks. Reviews begin with an assumption of a clean sheet to rebuild the curriculum from scratch, based on the needs of the national security strategist at a given point in history rather than simply to revalidate or tweak the existing curriculum. Once those charged with the curriculum review develop recommendations, the committee presents its recommendations of specific learning outcomes to measure the achievement of specific strategic goals in a coherent curriculum to the faculty as a whole along with the Commandant. The latter charges course directors to develop specific learning methodologies by which to achieve the curriculum outcomes for a 3-year term. This specificity of term allows for a constant and ongoing reappraisal of the needs, techniques, and emphases of the College and guards against staleness. Coupled with the ongoing peer review of individual core courses along with student feedback after the completion of a course, the core courses sustain significant regeneration regularly, commensurate with changes that strategists confront, including techniques, emphases, and relative priorities for the U.S. national security community.

Through the process of a curriculum review, conducted by a wide range of faculty from the military, agency, and Title X backgrounds, curriculum evolves in a painstaking manner. After detailed discussion among the curriculum committee and in consultation with the other faculty and College leadership, the committee develops a series of learning objectives that become closely associated with course objectives for the graduates of the College to understand upon completion of the academic year. The 2006 curriculum review included the following objectives, as noted at *www.ndu.edu/nwc.*

- National security strategy and policy are formulated and implemented within international and domestic political processes and environments that are dynamic, changing and replete with competing interests. As a consequence, policy is often as much an outcome of bureaucratic processes, compromise, and the influence of a dominant personality as it is of rational calculus.
- State resources are limited, requiring policymakers to set priorities among competing domestic and international interests, and to accommodate the allocation of resources between selected domestic and international objectives. Means and ends must be judiciously matched within strategies designed to accomplish national objectives.
- National security objectives and strategy must be devised and implemented within environments where ethical norms inform and constrain policymakers' freedom of action.
- A national security strategy must identify the interests of the nation and the challenges to those interests, and specify the objectives to be met through the use of specific policy instruments, particularly in any use of military force.

- Instruments of policy must be orchestrated within a cohesive strategy that deliberately integrates the selected instruments to achieve specified objectives.
- As a component of national security, military strategy and operations require the development within the Armed Forces of a joint culture that fosters the teamwork essential for deterrence, joint warfighting, and multinational endeavors. Planning and prosecution of joint campaigns and major operations require competency in joint skills, including the ability to orchestrate air, land, sea, space, and special operations forces into effective joint teams.

The core course directors develop courses to underscore these learning objectives, elective course directors develop courses to reinforce these issues, and all faculty seminar leaders reinforce these themes throughout the academic year.

BASIC PARAMETERS OF THE ACADEMIC YEAR

Students at the College do not live at Fort Lesley J. McNair and often commute more than 2 hours daily between their homes and the College. Each week has a maximum of 13 contact hours when a student can be in class with instructors, in a lecture, or in a structured, designated academic activity. This includes core and elective courses. The reason that the College has developed this limitation is the reality that students, confronted with the extreme commuting times in the Washington, DC, metropolitan area at the senior levels in their careers, with family obligations after the rapid operations tempo that today's military confronts, will not do the necessary and crucial readings (averaging 500 pages weekly) if they do not have sufficient time to read, digest, and apply the readings after their seminar learning periods. Before the curriculum reforms of the late 1980s, resulting from the congressional push towards civilian accreditation, contact hours crept up towards 20 hours weekly, but the student learning, evidenced by student theses and evaluations from students, appeared to stagnate due to insufficient time to think about the problems at hand. For the two 12-week periods of the academic year when students enroll in two electives minimum[5] per term, the core course contact hours are nine per week along with 4 hours for electives. When electives are not in session, the core may bump up to 13 hours in lecture or seminar weekly.

TEACHING NATIONAL SECURITY STRATEGY

The curriculum of the NWC consists of specific core courses and teaching techniques that permeate the academic year. Seventy percent of the students' time is spent in core curriculum, which has several components. Seminars, led by a faculty seminar leader, allow 13 students to meet three to four times weekly, often following lectures in plenary but not exclusively so. The students do not randomly choose the seminars, mates, or instructors. An algorithm distributes the 200-plus students[6] to craft a discussion experience with students from across the Services to exploit the range of expertise rather than allowing for concentrations of student specialties. Each seminar has two international fellows from different regions of the world, one Marine Corps representative, one person from the intelligence community, one State Department Foreign Service

Officer, two Navy or Coast Guard from differing branches, two Army, two Air Force, and two students from other civilian agencies. Seminars are the primary mechanism for learning, with support from in-house and external speakers who bring specific knowledge on a topic for all to understand and to mine pedagogically. These lectures in plenary are supportive of the seminars, not the other way around, and are only one method of educating the developing strategist.

The unified core curriculum generally transpires during the mornings 4 to 5 days weekly, although over the past 5 or 6 years Wednesdays have increasingly become the day for the National Defense University (the NWC's umbrella administrative institution) to host a Distinguished Lecture series. The Core, as the bulk of the curriculum is known, has seven component parts run consecutively throughout the academic year. Each student takes all seven parts, and everyone is enrolled in the unified core curriculum at the same time. The components of the Core are 6100 – **Introduction to Strategy**, where students learn about inquiry, the nature of strategies, comparative strategies, and strategic logic; 6200 – **War and Statecraft**, where students begin their appreciation of war as an instrument of statecraft as well as an influence on any society; 6300 – **Non-Military Elements of Statecraft**, where students ponder the instruments available to the nation beyond the blunt force of arms to achieve national security strategies; 6400 – **The Domestic Context and Decision-Making**, which may be most students' first exposure to the importance of the U.S. domestic context to the ability to achieve a national security goal and how this affects decisionmaking; 6500 – **The Global Security Arena**, which puts students into the shoes of others around the world by examining the national security concerns of states and actors outside of the United States along with exploring the emerging threats and opportunities confronting the United States; 6600 – **Applications of National Security Studies**, which gives students an opportunity to apply their learning from the year in three extended case studies (most recently alternative futures for the United States, nonproliferation and the current problems of Pakistan as a fragile state, and options for a strategy with China over the next 15 years); and 6700 – **Field Studies in National Security Studies**, which is a year-long exploration of U.S. strategy for a single nation or subregion of the world that also requires the students to validate their learning by studying on the ground in the chosen region or state.

Students also take a minimum of four elective courses designed to complement the individual's interests while supporting the core curriculum. The student's choices include courses at the NWC, the Industrial College of the Armed Forces, the Information Resources Management College, the Center for Technology & National Security Policy, and the College of International Security Affairs, all components of the National Defense University. The electives have a variety of formats and emphases, common only in their duration of 12 weeks per semester. Because of the variety of backgrounds that students bring, courses are introductory as well as allowing greater depth of academic study beyond the core. These electives allow students the opportunity to delve much more deeply into a subject, but equally give students an opportunity to study in some depth a wider range of topics than does the Research Fellowship discussed below. Electives complement the rest of the curriculum with the fundamental understanding of their role in supporting the development of the strategist.

The only constraint on the student's choice is that one of the courses must have a regional concentration to support the Field Studies component of the Core. Field Studies is an in-depth pursuit of understanding of a region or, in several cases, the U.S. national security relationship with one nation. Students choose between roughly 24 Field Studies opportunities, two of which are generally thematic and the remainder are overwhelmingly regional designations. The Field Study involves an elective designated to emphasize the importance of the country or region, along with more than a dozen academic preparation sessions to highlight specific concerns relative to that particular Study, such as experts coming in to discuss the economic interrelationship between China and the United States, the military deterioration of Russia, the role of Islamic law in states of the Middle East, or the role of ethnic and religious differences in central Africa, which round out the topics that the Field Studies students need to pursue in each study.

THE ROLE OF QUESTIONS AND FRAMEWORKS IN THE STUDY OF STRATEGY

The College proudly notes that its philosophy is to teach students how to think, not what to think. The College has never sought to have a single framework of analysis but to provide the students with the ability to analyze any problem in a rigorous, methodical manner. The College encourages its faculty to use frameworks as part of the teaching pedagogy but there is no single framework offered. The dominant approach, developed and still lovingly refined by Dr. Terry Deibel during his 30-plus year career at the NWC, now appears in his book, *Foreign Affairs Strategy: Logic for American Statecraft*.[7]

Deibel's model is a complex, interactive, dynamic one that reminds the analyst that ends-ways-means requires a reconciliation as much as a mathematical equation does. The pivotal value of Deibel's model along with the rest of those that the College curriculum supports is that the foundational assumption at work is a Clausewitzian one. Ends-ways-means does not refer exclusively to use of power in relation to a specific military challenge but includes serious weighing of nonmilitary elements of statecraft, the role of societal support or discontent, consideration of the context in which opponents and allies operate, along with recognizing that every single decision has consequences. The ends-ways-means equation implies that the strategist must reconcile the use of instruments with national interests within the context of what is acceptable to the U.S. body politic.

We should insert a caution about the ends-ways-means paradigm. Strategy is an adaptive and inclusive concept, always adjusting to circumstances, always subservient to the ends of policy but amenable to the changing nature of the opponent, to changing resources, and to timelines. Accordingly, there are no risk free strategies, an admonition which is unlike the precision implied by balancing end-ways-means. It is also a profoundly human enterprise. Leadership, personality, will, the vulnerabilities of small group dynamics, and culture matter. It is the interaction of these variables that affect the making, the implementation, and the results of strategy. Therefore, a serious challenge in the teaching of strategy at the NWC, and probably at all the war colleges, is to correct the student propensity to think only in terms of the application of kinetic resources. To prevent students from defaulting to a mechanistic calculation that emphasizes kinetics, the war colleges should teach that there must be a permanent conversation between policy and strategy, so that American power can be applied more effectively.

Several instructors have developed their own models over the years and are encouraged to continue doing so throughout the year. Several emphasize aspects of the strategic environment to a lesser or greater degree. Others start with a completely different set of assumptions before starting to do the analysis. One example is someone starting with the assumption that the United States has national interests only to defend its borders, thus an Iran with nuclear weapons or an Israel requiring U.S. support is not a national concern for the United States any more than a strong China is a threat. The role of assumptions takes the trajectory of analysis to vastly different strategic objectives depending upon the assumptions used.

Rigorous analysis requires students to begin applying a natural process of thinking that will encourage them to include all of the factors that a strategist must consider in crafting national security strategy. Experience in comparing incoming students with those departing after a course of study generally shows that those who come in without focusing on analytical thinking move immediately to the use of tools of statecraft without understanding that significant complications and contradictory effects may ensue if one does not start with understanding national interests, objectives, and the need for domestic support.

Student seminars, led by faculty seminar leaders (FSLs), are the primary vehicles for learning rather than faculty lecturing or self-study. While student learning requires considerable time to contemplate the 500 pages of weekly reading, the exchange of ideas, guided by faculty in an academic-freedom enhanced environment, is the focus of learning. The students with their careful mix of capabilities and experiences, due to the algorithm for seminar composition, can discuss material in greater depth. Student readings support each topic for each course, and each topic has targeted learning objectives and questions for consideration. Faculty interaction in the seminar aims to redirect discussion, where necessary, to meet the objectives. NWC students also have the responsibility to present materials for many case studies, run seminars on specific topics, and engage fully in their seminar experiences.

Faculty seminar roles involve asking provocative questions via the Socratic dialogue rather than imparting knowledge in the more traditional faculty-centric model of education, which uses the lecture as the principal pedagogical vehicle. In true seminar learning, the student evolves in understanding a topic and begins to develop his or her own appropriate questions about each topic. The point of a master's degree is not learning basic knowledge but learning to develop and explore the appropriate comparative tools of understanding in any particular setting. Student evaluations invariably heap the highest praise on those FSLs who are most effective at guiding with gentle hands rather than trying to impose their views on a seminar or a topic.

The questions that this author favors include those which remind the student to identify basic assumptions at all points in any analysis. Questions are most appropriately tailored to the individual case studies being used but types of questions, used to open students to wide-ranging implications, are crucial to students developing the facility to marshal the range of data available in any particular context to craft a strategy. In considering U.S. strategy towards Southwest Asia, for example, questions would include asking what the basic national interests are in the region, along with what the context is for the decision about to be taken. Another entire category of questions relate to what the options are for using various instruments of power and what consequences, intended and unintended, of each use would follow. This latter category is particularly important because the consequences of a strategic decision may not be apparent for years to come but need to be taken into consideration, nonetheless. The necessary questions to ask students in the safety of a seminar setting also include the specifics of how they would use any instrument and how that use would be received by the publics at home and aboard.

RESEARCH FELLOWS

The College no longer requires a thesis requirement because the quality of these works in the past was inferior to the learning that students gained from interacting in the seminar. While an exceptionally small number of students each year choose the Research Fellow option at the War College, their selection involves a rigorous review process. In the past, too many students came in with exceptionally vague agendas to pursue, often accompanied by shallow understanding of the specific research questions under study. Because of increased faculty scrutiny over research proposals, students who do choose the research fellow option and receive admission to the program have proven more successful in their work. During the first decade of this century, Research Fellows have averaged three out of a class of 223 annually.

Students who are Research Fellows enroll in the core curriculum, as do all other students in any given academic year. The Research Fellow receives 10 days in the academic schedule to miss seminar for work on the research project and is exempt from three of his/her four elective courses, thereby taking only the Field Studies-supporting elective for the appropriate geographic area to which the student will travel in May of the academic year.

CASE STUDIES

Case studies, made famous by law and business schools in the United States, offer a vital teaching tool at the College. Each core course uses case studies to open the door to students wrestling with material at the same time that the particular case highlights its importance for historical or policy reasons. The case study may be Thucydides and study of the Peloponnesian War, highlighting the linkages between strategy and public opinion, or ends-way-means; use of irregular warfare in Afghanistan to indicate the complexity of operations; or President Abraham Lincoln as a wartime leader to indicate the importance of any individual in a wartime setting. Each core course and many of the more than 48 electives offered each year use case studies to highlight specific points for the national security strategist to consider.

Case studies may be historical in nature, such as the use of the case studies produced at the College by its faculty or students, or the commercial products written at other academic institutions. Similarly, case studies may be hypothetical for learning purposes such as a nuclear nonproliferation topic looking at the implications and strategic options for an Iranian nuclear device or program. Case studies, all of which have been used in the "Applications in National Security" course for the past 2 years, may also look at a broad phenomenon not as traditionally linked to national security, such as the problem of a pandemic outbreak for strategists.

Case studies may emphasize instruments of statecraft, elements of national security such as the domestic, foreign or global actors involved in any particular scenario, or they may look at interagency concerns at work. Each case study is crafted specifically to meet the learning objectives of the particular course. Case studies vary from year to year according to the choices of the individual course directors, as each tries to implement the College's learning objectives as well as according due weight to the global context, which offers targets of opportunity for relevant cases.

STUDENT WRITING REQUIREMENTS

One of the most hotly debated subjects for the faculty is the appropriate approach and amount of writing that students use at the College during their 10-month program. One school of thought argues that students need to do expanded writing assignments (minimally 30 pages per topic) with a theoretical base as would occur in a traditional graduate school. These are broader, sweeping questions that would cause the student to pursue the work across the academic year or at least for a longer period than the usual 5- to 7-week core course. This would force students to think about the application of a comparative and theoretical base to a particular problem confronting strategists. The second school of thought argues that no graduates of the College will regularly use this approach to writing since the most common format that graduates will use in future assignments is shorter, analytic papers on subjects related to individual courses in which they are enrolled.

In the 1980s, the College used masters' theses for its students, but that approach ended around 1990 as the curriculum tightened up. Faculty saw evidence that the theses were not of a satisfactory quality. Instead, the College developed 7- to 12-page papers for each core and most elective courses. The papers match with the course objectives but use a variety of formats. One is strictly a research paper, requiring interviewing on a policy question with which the student has personal knowledge. Other papers relate to the strategy context, the use of a nonmilitary element of statecraft in a particular case, the use of the military instrument of statecraft in a particular military campaign, and similar approaches for the four electives in which the student enrolls. Occasionally, an elective will have an oral or group project in place of the writing assignment, but those circumstances are rare because the College believes that writing forces rigorous thinking beyond the seminar.

ELECTIVES

Some of the most innovative thinking about strategy at the College transpires in elective courses. Students enrolled at the College have the option of requesting elective courses at the Industrial College of the Armed Forces, Information Management Resources College, the College of International Security Affairs, and the Center for Technology and National Security Policy, all components of the National Defense University at Fort McNair. Courses are all 12 weeks in duration and 2 hours per meeting; and run the gamut in subject matter, techniques, evaluation processes, and learning outcomes, except for the broad overall goal of supporting the students' learning in areas of national security strategy. Courses may be thematic or regional in nature, historical case studies, or a range of other ideas. Students at the National War College must enroll in two electives per term with the sole restriction that one of their four annual courses must support his/her Field Study in 6700 – **Field Studies in National Security Studies**.

These courses allow faculty to follow their interests and to test drive new techniques, themes, and approaches. The University offers well over 100 courses per term on a range of topics. One of the most innovative is the two-term course, "Warriors at Battle," which is an interagency, future problem, lead-in to war gaming, which has historically prepared the student for the Joint Air Land Sea Service war game in the spring of each academic year at Maxwell Air Force Base in Montgomery, Alabama, where NWC students play against other senior service college peers.

NONTRADITIONAL LEARNING VEHICLES FOR NATIONAL WAR COLLEGE STUDENTS

Additionally, students do considerable national security strategy learning during their interactions in committees, a structure roughly equating to a home room. The educational value of this system is that students in a particular committee are distributed carefully to preclude the concentration of any particular type of students from any Service or specialty in the committees. The committee comp-osition remains constant throughout the year, allowing for the development of friendships that lower professional barriers to facilitate more candid, sustained discussions without the professional jealousies that single opportunities to present a view may generate. Committee discussions also create a bonding that allows long-term friendship and respect to grow among people who would have had no other mechanism by which to engage in a sustainable exchange of knowledge with and about other leaders of the Services. These government leaders will likely come into contact with each other in the future at higher ranks.

The College approach to strategy is thus integrative and holistic. It does not rely overly on any single aspect to teaching strategy but tries to reinforce the need for systematic, rigorous thinking for all students (and faculty). The function of the various portions of the curriculum is to allow a method of thinking to solve the exploding array of strategy challenges confronting the new national security senior leadership in the United States and around the world. Students learn from their extensive period in traditional lectures in plenary or in seminars of 13 people, but that is not the only mechanism by which the students learn. In addition, the College utilizes, through its careful attempt to maximize the interaction among students of so many different experiences across the security community, as many *fora* for discussion and understanding as possible. Oral evaluations give strong indications of the success of the approach although consistent and constant reappraisal of the program forces the College to recalculate the specific topics, themes, and techniques regularly. As noted to a Board of Visitors' inquiry about how the College keeps its faculty current on materials for teaching, one faculty member noted

"Anyone who does not keep up with strategic concerns will be savaged by these people who understand the real issues at work" in today's national security strategy requirements.[8]

FINAL OBSERVATIONS

No one at the National War College is arrogant enough to believe that the process of educating strategists cannot be improved; it can and should be. One of the most interesting aspects to conceptualizing the curriculum and implementing national security strategy education at the NWC is the constant reappraisal of the subject matter. While improvements are always welcome, the process of evaluating content, pedagogical techniques, readings, speakers, educational technology, and outcomes of teaching is ongoing to provide the most current information and to put those outcomes within the changing international context against which all strategies must be applied.

It would be an ideal world if students had the luxury of a 3 or 4 year period to study the international context, the use of Special Forces or irregular warfare, or the budget process, but today's operational tempo does not allow such a luxury, not even a 2-year effort. The absolute zero sum nature of time away from their Services means that something added to the curriculum requires that something else be deleted. The effect is that some topics cannot be covered, to the great disappointment of all. But the constant reappraisal of the curriculum allows for consideration of what the highest priorities are at the beginning of any academic year.

Strategy decisions in the United States are not the exclusive purview of the military or of the national security community writ large. The President of the United States and his civilian leadership, in consultation with professionals in diplomacy, defense, homeland security, justice, and the other departments and fields along with appropriate advice from the legislative branch share such responsibilities, hopefully producing judicious decisions. Rarely, if ever, do civilian elected leaders or political appointees of the executive branch or legislature have the opportunity to study strategy in a formal setting such as the NWC or any of the several senior service colleges, or even traditional academic institutions (where such courses are rare) that focus on this topic. The trajectory of civilian leadership into strategy making is a fundamentally different one.

To this author, the most fundamental aspect to teaching strategy, however, is the requirement to help those conceptualizing and implementing to ask appropriate questions as much as to have them filled with certain knowledge. The ability to ask questions, even if they appear well outside the norm, may prevent the types of failures in strategic thinking that so many critics accuse the United States of making over the past 20 years. The importance of reevaluating assumptions and remaining open to different interpretations of assumptions is utterly crucial. The questions may relate to allies, economic balances, the role of nation-building or of prior similar conditions, or a range of issues. The purpose of the questions is to put the strategists in the most open position to examine all of the options to achieve the national goals. While two situations will never be precisely the same, the openness to understanding the similarities allows the strategist to think about the most appropriate approach in using the elements of statecraft available. The assumptions and questions that the strategist asks will help to determine the answers to those questions. In today's world, these analytical explorations need to be as wide-ranging as possible while remembering that there are views held by others around the world that are likely to differ strongly from those of the United States and that there are national interests elsewhere that on occasion will clash with or offer support to the interests of the United States.

On a more personal note, the author prefers emphasizing the role of context and assumptions as the dominant pedagogical approach to others used at the College. Knowing that students recognize concerns specific to any particular nation or perhaps a problematic region is fundamental to crafting a strategy appropriate to the case. While models are important, they only fit relative to the flexibility that they allow for the specific instance where they are deployed. Since the context and assumptions in any case may change because of conditions on the ground or as a result of adjustments made by government, insurgents, international actors, or any other player in the relevant case, understanding and appreciating the specifics of the case seems more important than any other single aspect to the case. The NWC has been known since the 1960s for emphasizing area studies, a holistic consideration of language, history, population characteristics, governance challenges, economic weaknesses and strengths, among other things. This attempt to use an interdisciplinary appreciation for what is going on in any particular part of the world is not an attempt to make NWC students into experts on South Asia, for example, but gives them an understanding that members of the diverse community of nations have their own basic considerations for strategic problems upon which they will firmly fix their analysis, and their own strategic considerations to which they will cleave as strongly as we do to liberty, freedom, and other basic tenets. Without this understanding, War College students may flounder as they attempt to be national security strategists in the years to come.

ENDNOTES - CHAPTER 6

1. Mission Statement of National War College, available from *www.ndu.edu/nwc/index.htm*.

2. A starting point for reference on the NWC teaching of strategy is the College webpage, available from *www.ndu.edu/nwc*.

3. The Department of State has held a pivotal role in the College since its inception in the 1940s. Additionally, for decades State Department contingents were the single largest component of the non-DoD students since the College's mission was to educate future leaders from the civilian and military who were going on to national security leadership. The State role has been diminished somewhat after the Cold War, and especially after the greater emphasis on homeland security since September 11, 2001, as students from other agencies have become included in the incoming classes.

4. This is the term for the war colleges for the Army, Air Force, Navy, and Marine Corps, along with the Industrial College of the Armed Forces. These are all open to students at the upper end of their professional military education experience in the uniformed Services. The term "senior service schools" is also used interchangeably.

5. A small number of students take more than two electives in any given term.

6. In the post-9/11 world, most classes at the NWC have had roughly 225 students. The size of historic Roosevelt Hall, home

of the College, restricts the class sizes to no more than 239 to accommodate students, faculty, and the few administrative staff members who help carry out the academic year at the College.

7. Terry L. Deibel, *Foreign Affairs Strategy: Logic for American Statecraft*, Cambridge, MA: Cambridge University Press, 2007.

8. Author's response to a Board of Visitors' member's question on May 28, 2009, at Fort McNair.

CHAPTER 7

HOW DO STUDENTS LEARN STRATEGY? THOUGHTS ON THE U.S. ARMY WAR COLLEGE PEDAGOGY OF STRATEGY

Harry R. Yarger

The proper practice of strategy is the most significant challenge confronting senior leadership in the American military today. Recent experience has raised questions about the strategic competency of our leadership and the state of the national security profession. Consequently, how students learn strategy is a serious question for the military profession. It begs the questions of what we mean when we say strategy and how those who profess to be strategy faculty at the senior service colleges (SSCs) should teach it. This chapter examines how strategy is taught at the U.S. Army War College (USAWC) and how students are expected to learn.

We know what the nature of strategy is because experts from Sun Tzu to Clausewitz to Colin Gray have told us in their own ways that strategy is both the science and art of practicing the profession at the highest levels. Sun Tzu admonished others to both heed his plans (rules) and seek victory from the situation through understanding of the mutually reproductive forces of *cheng* (normal or direct) and ch'i (extraordinary or indirect).[1] Clausewitz follows a similar path, recognizing in his examination of war and strategy that science is knowledge and art is creative ability, but placing war itself in neither realm, but rather identifying it as a part of man's social existence.[2]

Nonetheless he concludes:

> This much is clear: this subject, like any other that does not surpass man's intellectual capacity, can be elucidated by an inquiring mind, and its internal structure can to some degree be revealed. That alone is enough to turn the concept of theory into reality.[3]

Hence, art and science together transcend either singularly and both have a role in war and strategy. Colin S. Gray deals in detail with this duality in *Modern Strategy*, a book that applies social and political science to explain the timeless aspects of the nature and function of strategy in modern terms, so that others may learn how to practice the art.[4] Since these theorists wrestled so hard with understanding strategy, it should come as no surprise to us that strategy remains difficult to teach and to learn. So what does a professional educational institution like the USAWC do to teach this critical subject and what should it tell its students so they can learn the practice of strategy.

PURPOSE IN TEACHING STRATEGY

To determine what it should do to teach strategy, the USAWC first had to answer three questions. The first of these is about purpose: *Why are we teaching strategy in the curriculum?* At first glance, this appears to be obvious—to educate strategists. On closer examination, it is less obvious. As its mission statement makes clear, the USAWC's focus is on strategic leaders: "USAWC prepares selected military, civilian, and international leaders for the responsibilities of strategic leadership in a joint, interagency, intergovernmental, and multinational environment."[5] Therefore strategy as a subject at the USAWC is logically viewed through the lens of strategic leadership. Equally important, strategy within the profession is about much more than being a strategist even though the pure strategist is instrumental in the profession's successful practice. Within the profession, strategy is about being able to formulate strategy, understand it, critique it, promulgate it, and execute it through appropriate planning and tactics — and in this light strategy is perceived culturally by the USAWC as the lynchpin of leadership at the strategic level. It marks the distinctiveness of strategic leadership defining the environment and level of skills and competencies; the realm of military advice; and the bridging mechanism between policy and the military in war.

Major General Richard A. Chilcoat and a task force of USAWC faculty members wrestled with the complexity of the roles of the complete strategist in an insightful monograph entitled *Strategic Art: The New Discipline for 21st Century Leaders,* published in 1995. Their conclusions are part of the teaching culture of USAWC even today and affect the pedagogy. In their judgment, strategic leadership is the effective practice of the strategic art. Strategic art was defined as the skillful formulation, coordination, and application of ends, ways, and means to promote and defend the national interests.[6] A master of the strategic art had mastered the three roles of: strategic leader, strategic theorist, and strategic practitioner. The first centers on the abilities to provide vision and focus, capitalize on command and peer leadership skills, and inspire others to think and act. The second centers on the abilities to study the history of warfare, derive relevant insights, formulate strategic concepts and theories, integrate these with the elements of power and national strategies, and teach and mentor in regard to the strategic art. The third, the strategic practitioner, centers on the abilities to deeply comprehend the levels of war and their relationships with strategy; develop and execute strategic plans derived from interagency and joint guidance; employ force and other aspects of military power; and unify military and nonmilitary activities toward common objectives. Each of the roles has distinct skills and competencies, but they also share common ones. While the master is competent in all three, in practice different personalities, positions, and environments may make one of the roles dominant. (See Figure 1.)

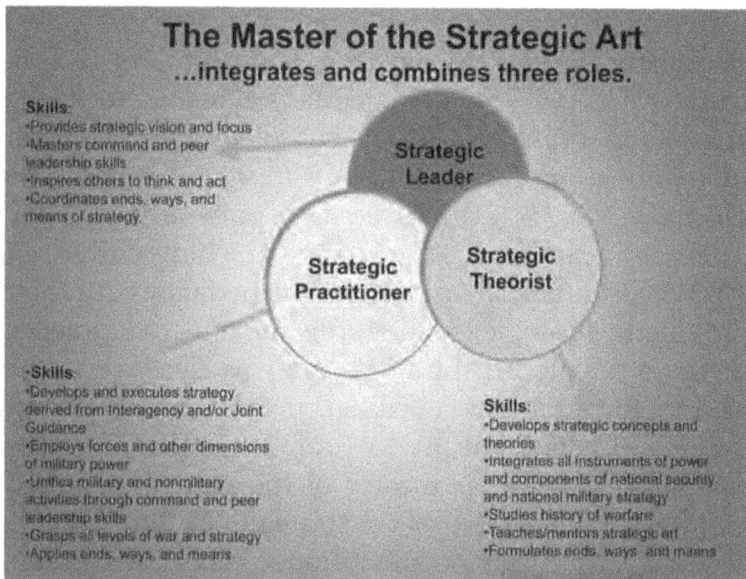

The Master of the Strategic Art
...integrates and combines three roles.

Skills:
•Provides strategic vision and focus
•Masters command and peer leadership skills
•Inspires others to think and act
•Coordinates ends, ways, and means of strategy.

Strategic Leader

Strategic Practitioner

Strategic Theorist

•Skills:
•Develops and executes strategy derived from Interagency and/or Joint Guidance
•Employs forces and other dimensions of military power
•Unifies military and nonmilitary activities through command and peer leadership skills
•Grasps all levels of war and strategy
•Applies ends, ways, and means.

Skills:
•Develops strategic concepts and theories
•Integrates all instruments of power and components of national security and national military strategy
•Studies history of warfare
•Teaches/mentors strategic art
•Formulates ends, ways, and means

Figure 1.
Each of the roles has distinct skills and
competencies, but they also share common ones.

In their search for what mastery meant, the task force acknowledged that varying degrees of competence and emphasis in the roles were not only probable but were even relatively desirable based on the professional position and circumstances.[7] Nonetheless, their explanation recognizes in each part the same seemingly contradiction articulated by Sun Tzu, Clausewitz, and Gray — strategy is both a science and an art. Any pedagogical approach and any student learning must consider both — singularly and together. Each is essential to the profession, but the art is more difficult to learn and, consequently, to teach. To its credit, the USAWC attempts to do both in educating leaders for the strategic level, perhaps being more pedagogically clear in teaching the science than the art and thus making the science the logical place to start in a discussion of learning strategy.

WHAT TO TEACH

The second big question is: *What should be taught?* Within USAWC this question is answered initially through a scientific approach — one of theory, processes, and models. The very notion of a **science** of strategy suggests that we can study strategy formulation, theorize about it, and improve performance by better understanding the processes involved. Clausewitz properly articulates the purpose of science in regard to strategy ". . . these principles and rules are intended to provide a thinking man with a frame of reference. . . ."[8] Within the USAWC, faculty members have been particularly productive in providing a science of strategy for student learning. In the early 1980s, USAWC faculty members were advancing key ideas on both strategic leadership and strategy. Colonel Charles A. Beitz and the faculty of the leadership department created a minor renaissance in leadership studies by their focus on the uniqueness of senior leadership skills and competencies, drawing on business and political models as well as military ones. A similar renaissance in strategy was in progress at that time with the introduction of the works of Clausewitz, popularized in part by U.S. Army Colonel Harry Summer's *On Strategy* published in 1981.[9] Taken together, these two minor renaissances founded the modern USAWC cultural perspective of strategy as articulated in Chilcoat's monograph.

In the USAWC teaching culture, strategic leadership and strategy are nearly synonymous. Leadership at the strategic level requires a unique set of skills, competencies, and knowledge that enables the leader to manage the volatile, uncertain, complex and ambiguous (VUCA) strategic environment. The science helps the student to understand the nature of the environment, the strategic thought process, the skills and competencies required, and the processes of the national and international systems. Faculty members at the USAWC take the role of science seriously because it creates a common basis for understanding and discussing strategic issues and strategy. For example, the USAWC models strategic thinking based on theory and scientific research in various fields, advancing the idea that formulation and articulation of strategy is supported and enhanced by scientific understanding of thinking skills and competencies. (See Figure 2).

Late in the 1980's Art Lykke's three-legged stool model captured the essence of a rationally stated strategy in its ends, ways, means, and risk paradigm.[10] From this early start, a rich literature of articles and monographs that examine the theory and practice of strategy were produced by USAWC faculty, including Dave Jablonsky, Steve Metz, Robin Dorff, and numerous others — and this flow continues today in a collected works strategy anthology published biannually.[11]

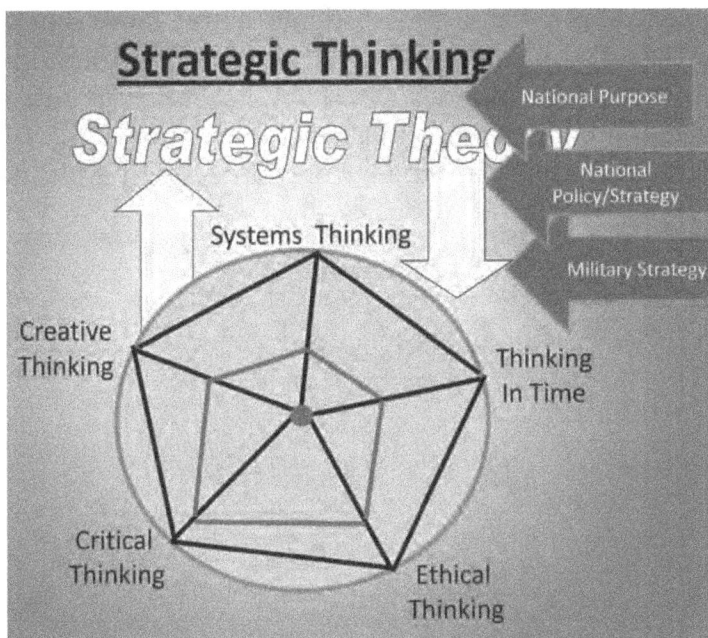

Figure 2.
Faculty members at USAWC take the role of science seriously because it creates a common basis for understanding and discussing strategic issues and strategy.

In response to a challenge laid down by Gregory D. Foster in 1990 for the articulation of a general theory of strategy, this author proposed such a theory in 2006 based largely on an expanded synthesis of what has been taught on strategy at the USAWC for the past 20 years.[12] (See Figure 3.)

The science of strategy is also evident in the current USAWC Strategy Formulation Model, which integrates strategy formulation into the context of the strategic environment and policy formulation. This model resulted from the work of multiple faculty members over a number of years. It is a graphic representation of the content studied in two of the core courses—and helps the students better comprehend the very complex process of U.S. strategy formulation. In a similar manner, USAWC faculty members developed variations on strategic appraisal models to assist students in the learning process. In addition,

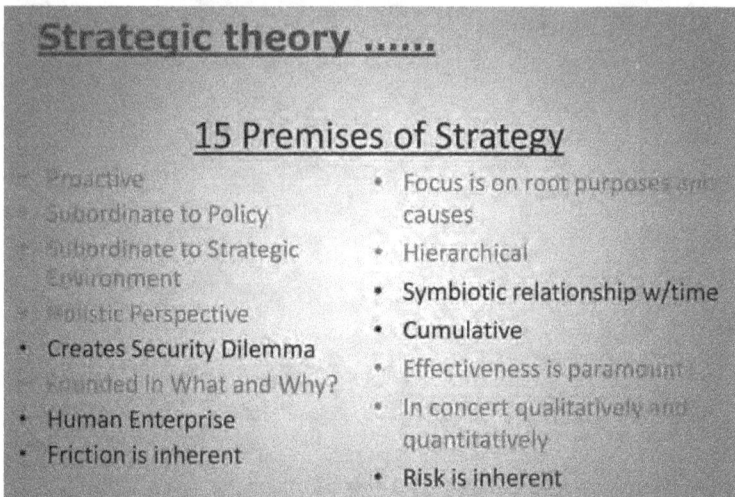

Figure 3.
... a theory ... based largely on an expanded synthesis of what has been taught on strategy at USAWC for the past 20 years.

Sun Tzu's *The Art of War*, Carl von Clausewitz' *On War*, Colin S. Gray's *Modern Strategy*, and many more classic and modern works lay a basis for a scientific understanding of definitions, theory, and processes related to strategy—all of which discipline strategic thought and provide for deeper thinking about and discussion of strategy.[13] USAWC uses the works of the masters and the writing of faculty at other institutions as well as practitioners in national security—but the work of USAWC's own faculty drives the pedagogy. (See Figure 4.)

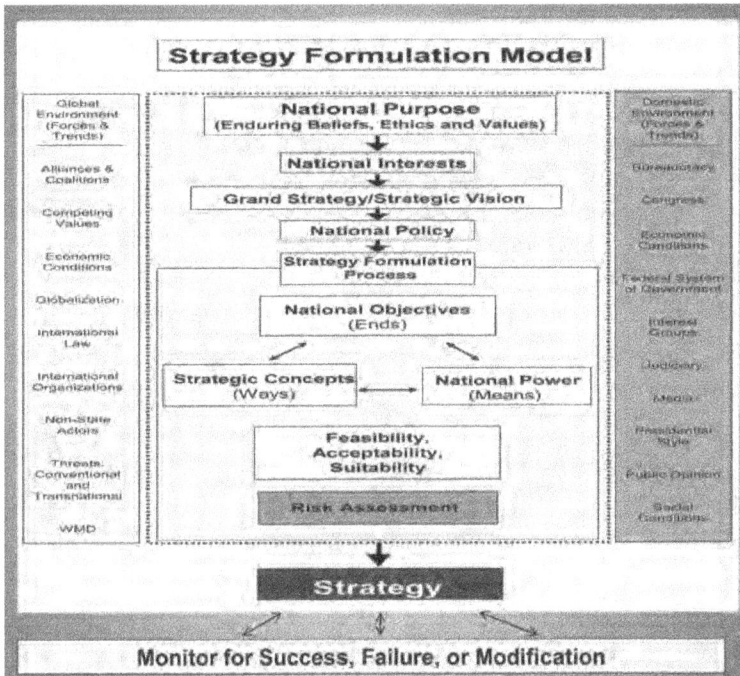

Figure 4.
The science of strategy is also evident in the current USAWC Strategy Formulation Model, which integrates strategy formulation into the context of the strategic environment and policy formulation.

The art of strategy is more difficult to learn. In fact some argue that it cannot be learned or taught, that true strategists are born. Such disagreement begs the question of what the art of strategy is. Clausewitz defined it as the ability to perceive what is important and to act in regard to it appropriately.[14]

> At this point, then, intellectual activity leaves the field of the exact sciences of logic and mathematics. It then becomes an art in the broadest meaning of the term — the faculty of using judgment to detect the most

important and decisive elements in the vast array
of facts and situations. Undoubtedly this power of
judgment consists to a greater or lesser degree in the
intuitive comparison of all the factors and attendant
circumstances; what is remote and secondary is
at once dismissed while the most pressing and
important points are identified with greater speed
than could be done by strictly logical deduction.[15]

We all know intuitively what the art is—simply
stated, it is the ability to see the strategic dots and
connect them in a meaningful manner to service U.S.
interests—to comprehend the strategy factors and
use or address them appropriately in the context of
the strategic situation over time. For Clausewitz, this
ability can only be ". . . attained by a special talent,
through the medium of reflection, study, and thought:
an intellectual instinct which extracts the essence from
the phenomena of life, as a bee sucks honey from
a flower."[16] For those less poetic, Clausewitz also
provides these other insights to how art is obtained,
arguing knowledge and theory ". . . must be so
absorbed into the mind that it also ceases to exist in
a separate, objective way. . . . [It] . . . will be sufficient
if it helps the commander acquire those insights that,
once absorbed into his way of thinking, will smooth
and protect his progress, and never force him to
abandon his convictions for the sake of any objective
fact."[17] Clausewitz' insights into the art of strategy are
reinforced by modern study as reflected in William
Duggan's *Strategic Intuition*.[18]

Although addressing strategy from a business
perspective, Duggan makes a compelling argument
that strategic intuition constitutes the art of strategy.[19]
It is distinct from ordinary intuition (gut feeling)
and expertise intuition (snap judgments). It takes a
prepared mind that understands the profession and

has studied the data in regard to the issues (Plod). It requires a lot of time understanding your relationship to the issues in regard to your purpose, desired outcomes, and objectives (Plot). It also requires you to free your mind to accept new ways of thinking and relationships (Play). If Plod, Plot, and Play are done at the same moment, the brain makes something new out of the data and yields a strategy — the insight that distinguishes strategic genius. Duggan's work is founded in both modern advances in brain science and historical experience. His scientific analysis is strengthened by his use of Clausewitz to support his thesis.

> Clausewitz gives us four steps. First, you take in "examples from history" throughout your life and put them on the shelves of your brain. Study can help, by putting more there. Second comes "presence of mind," where you free your brain of all preconceptions about what problem you're solving and what solution might work. Third comes the flash of insight itself. Clausewitz called it coup d'oeil, which is French for "glance." In a flash, a new combination of examples from history fly off the shelves of your brain and connect. Fourth comes "resolution," or determination, where you not only say to yourself, "I see!", but also, "I'll do it!"[20]

Duggan suggests that by understanding these phenomena, people can improve their strategic intuition.[21] This is not the same as arguing that everyone can be a strategic genius or a strategist proper. However, if we see strategy as inherent to the profession then we might conclude that members of the profession, even if few are strategists proper, should at least develop the level of art to the point that they can understand what the strategist's dots are, and how they connect to the extent of questioning the linkages and being able to plan the implementation of and execute the strategy. Duggan's and Clausewitz' insights on art are of value to the pedagogy of educating the profession as well as the strategist proper. (See Figure 5.)

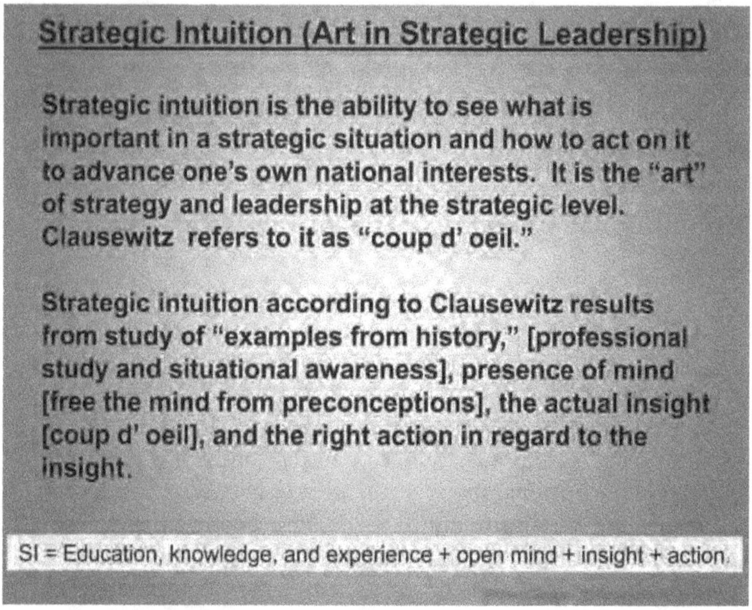

Strategic Intuition (Art in Strategic Leadership)

Strategic intuition is the ability to see what is important in a strategic situation and how to act on it to advance one's own national interests. It is the "art" of strategy and leadership at the strategic level. Clausewitz refers to it as "coup d' oeil."

Strategic intuition according to Clausewitz results from study of "examples from history," [professional study and situational awareness], presence of mind [free the mind from preconceptions], the actual insight [coup d' oeil], and the right action in regard to the insight.

SI = Education, knowledge, and experience + open mind + insight + action.

Figure 5.
...members of the profession, even if few are strategists proper, should at least develop the level of art to the point that they can understand what the strategist's dots are and how they connect to the extent of questioning the linkages and being able to plan the implementation of and execute the strategy.

Hilary Austen Johnson approaches this matter of art in a 2007 article in the *Journal of Business Strategy* in which she captures the contradiction, or paradox, of science and art in an educationally meaningful manner.[22] In her argument, artistry is emergent based on the ability of the individual to structure personal knowledge in such manner that he has mastered orthodoxy — applying developed knowledge — and is open to originality — generating new knowledge. In her model, mastery is concerned with recognition, effectiveness, skill, purpose, and focus. Originality is concerned with perception, creativity, innovation, flexibility, and openness. In true art, these two are successfully integrated in a dynamic balance of exploration (originality) and exploitation (mastery). However, in general, practitioners show a decided preference for one or the other and for good reason. Too orthodox, and practitioners avoid both risk and competitive advantage. Too original or enamored with novelty, and the lack of skillfulness or appropriateness in application raises the risk of short term failure exponentially.

Johnson argues personal knowledge must be structured on three levels to permit the emergence of artistry.[23] Directional knowledge is concerned with identity and motivation. Conceptual knowledge is concerned with understanding and organization. Experiential knowledge is concerned with sensitivity and skill. These are interrelated. Directional knowledge guides conceptual knowledge which in turn guides experiential knowledge. At the same time, experiential knowledge informs conceptual knowledge which in turn informs directional knowledge. Johnson's thesis is strengthened in this author's judgment because nothing she says contradicts what theorists of war and military strategy have said over the ages.[24] It is also supported by Johnson's articulation of three misconceptions related to learning and strategic art that are validated in other literature[25] and the author's own observation and experience in regard to strategy and art. The first misconception is that artistry is a matter of copying past success. In fact, strategic artistry is emergent—you cannot produce or predict strategic genius. This is not to say that educators cannot create the conditions that allow it to bloom. A second is that artistry has a final destination. In fact, neither strategy nor artistry has a final destination. Success simply creates new opportunities and challenges as others respond and conditions change. The third misconception is that achieving artistry assures success. In fact, artistry in strategy makes you a potentially better competitor, but neither assures success nor necessarily lessens risk. Acceptance of risk is ultimately a leadership decision based on the value of the gain in regard to costs and consequences.[26] Johnson's conceptualization of artistry reinforces the USAWC pedagogy.

HOW TO TEACH

The third question is: *How does USAWC teach strategy?* Our own historical experience, reinforced by adult educational theory, makes USAWC advocates of the seminar model of education at this level. This adherence to the seminar model has implications for faculty and students that is explored later. The structure of the current curriculum flows from both the USAWC leadership focus and its collective understanding of what should be taught in regard to strategy and the profession. In this curriculum, leadership and strategy are intertwined as are science and art. (See Figure 6.) In regard to the latter, USAWC is in good company as *The Art of War* and *On War* also recognize the relationship between science and art.

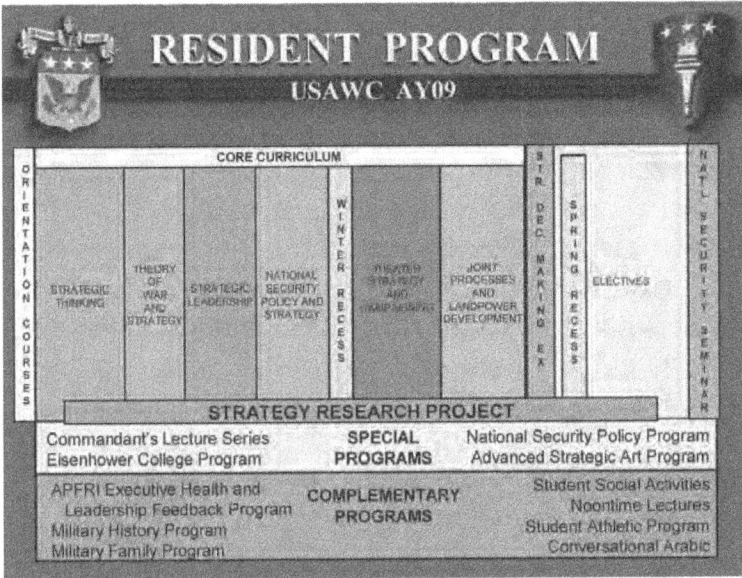

Figure 6.

... USAWC is in good company as The Art of War
and On War also recognize the relationship between
science and art.

In the USAWC model, the commitment to the art
of strategy comes together in the study of history,
case studies, current and past strategies, strategic
issues, regional studies, exercises, field trips, and
other practicum. In Johnson's model, this corresponds
to experiential knowledge, whereas the USAWC's
science would be conceptual knowledge and the
emphasis is on professional directional knowledge.
Few faculty members are under the illusion that
these applications do any more than contribute to
mastery — the emergent nature of artistry in strategy
is generally accepted, as opposed to planning, where
cause and effect hold sway. In the curriculum, these
applications are integrated into all the core courses, as
well as elective courses and other academic activities.

Sometimes these coincide in the same instruction as what might be termed science, and sometimes they stand alone as distinct case studies, exercises, and experiences. Taken together, as illustrated in Figure 7, the science and art presented at USAWC constitute the what and how of strategy teaching at USAWC in large measure, with the application turning inside the presentation of the science, sometimes contributing to the understanding of the concepts and sometimes practicing their application.

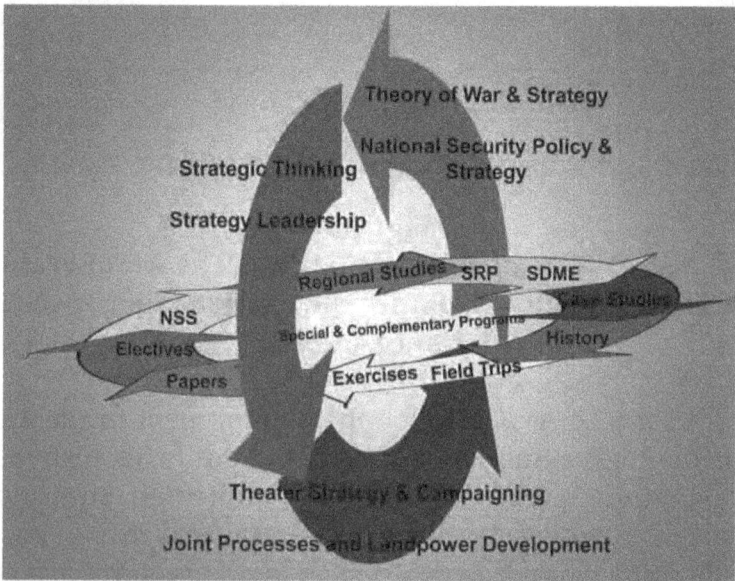

Figure 7.
... with the application turning inside the presentation of the science, sometimes contributing to the understanding of the concepts and sometimes practicing their application.

Within USAWC, as in other SSCs, we make our environmental scans seeking to understand today's issues and anticipate tomorrow's. We ask ourselves what these mean for us as an institution in regard to the profession. We critique ourselves and consider what our students accomplish and say. We consider the views of accrediting activities, outside critics, fellow SSCs, and the academic disciplines and the perspective of experience provided by our graduates and other practicing professionals. We review and revise our mission and our institutional learning objectives when it appears appropriate. Routinely we debate lesson subjects and content, their order in courses, and the number and sequence of courses in the curriculum. Daily we argue among ourselves about what leadership, policy, strategy, war, and the profession mean and how best to assist our students to learn. All of this is open to review and change—sometimes as part of a renaissance in internal thinking and sometimes driven by external events. Yet in the vein of Sun Tzu, Clausewitz, and more modern theorists, strategy remains for us both a science and an art, unequivocally linked to strategic leadership and the basis of professional advice in regard to policy and war.

While there may be—and should be—continuing debate in regard to specific content, sequencing, or the balance among methodologies, the basic pedagogy at USAWC is sound—founded in both educational theory and practice and the science and art of strategy, conveying directional, conceptual, and experiential knowledge in regard to the latter. It also supports the multiple needs of the national security profession in regard to roles anticipated for graduates. In large part, the disagreements we have among ourselves and with our external critics are ones of preferences and opinions—rarely of definitive substance and more in the realm of good and better ideas. Success in strategy education at institutions remains faculty and student centric. Thus, it can be rightly argued that if we want improvement in the quality of our teaching and learning—and hence our graduates' performance of the art—any serious advancement in how students should learn strategy likely lies in examining the responsibilities of faculty and students in creating and maximizing the learning environment.

THE RESPONSIBILITIES OF FACULTY AND STUDENTS

In any teaching institution, the role of the faculty is paramount and the high turnover of faculty inherent to any military schoolhouse notwithstanding, the faculty is the center of gravity—"the hub of all power and movement, on which everything depends."[27] The responsibilities the faculty members assume are instrumental to successful student learning. There are many ways of expressing faculty responsibilities. In the USAWC approach to strategic studies, three responsibilities are particularly important. First, strategy by its nature is multidisciplined. Consequently, faculty members must prepare themselves appropriately in all of these disciplines. Faculty can play to their academic strength but must seek to develop in their students the interdisciplinary intellectual capacity required of strategy. Second, because of this inherently multidisciplinary nature of strategy, faculty members must bring their own particular expertise into the curriculum and lesson development processes. Doing this means contributing to both the design and content of lessons and courses. The latter includes writing and developing of resources specifically for the USAWC's educational purposes.

Third, educational theorists have documented that students have preferences in learning styles. While the field has labeled these preferences in different ways and from different perspectives, in simple terms some are auditory, some visual, and some kinesthetic. Since students have various learning styles and will serve in differing strategic roles in subsequent assignments, faculty members must be adaptable in their teaching and facilitation styles to fit the individual learning style of each student and the collective mentality of the seminar. Furthermore, faculty members must approach these responsibilities with a developed appreciation of the value of each and a commitment to successfully integrate the three responsibilities into a coherent whole for the profession, the institution, the seminar, and the individual students. In this manner, the faculty members guide both the individual graduates and the profession as a whole to greater strategy competence — facilitating the development of the strategic art and its successful implementation.

Educational theorists also know that for the best education, adult students have to accept responsibility for their own learning. They inform us that students are more willing to learn when the relevance of the learning is clear to them — usually interpreted to mean they see an immediate use for the learning. Yet, students pursuing professional education, and particularly within higher military education, are asked to learn cognitive knowledge, concepts, and competencies; education that focuses on how to think as opposed to specifically what to think. Unable to predict exactly what students will be confronted with in their future career but knowing the nature of it, USAWC, like the other SSCs, prepares the student for the volatility, uncertainty, complexity, and ambiguity of the strategic level so that they can influence and manage the specific circumstances of their future environment. Consequently, the first responsibility of learning for students at this level is to accept that as a member of their profession they must understand their profession and seek out the knowledge, skills, and competencies required for their future roles. No one can predict with accuracy exactly what the roles or the circumstances will be for any particular student. What is known for certain is thinking and acting strategically is inherent to the practice of the profession at the strategic level. Faculty members can assist students in understanding their profession and facilitate their search for knowledge, skills, and competencies, but only the student can acquire them and practice them. Understanding their profession and accepting its obligations are the first, and the most important, learning responsibility of the student within a profession—it provides the relevance of learning and acknowledges the individual and collective accountability for learning.

Once this fundamental professional responsibility is accepted, the student must act on it. The second responsibility of learning at the USAWC is an individual commitment by the student to adequately prepare himself and to participate actively in seminar learning—including reflecting on the learning, revisiting key aspects as necessary, and practicing application as the curriculum unfolds. The third responsibility in the USAWC model is a collective responsibility. As members of a profession, students are collectively responsible for the learning environment. The USAWC embraces the philosophy of J. F. C. Fuller " . . . we shall teach each other; . . . it is only through free criticism of each other's ideas that truth can be thrashed out. . . . until you learn to teach yourselves, you will never be taught by others."[28] Experienced faculty know few things create better learning than preparing oneself for and teaching others. As members of a profession, students are collectively responsible for ensuring a successful learning environment. At the USAWC this means more than the individual responsibility addressed above, it implies a willingness to listen to and assist others. It requires the students to act in faculty roles as well as enforce seminar norms that promote quality learning.

A fourth learning responsibility for students is the duty to reflect on the lesson, course, and curriculum designs, and critique them meaningfully. Meaningful critique is not about whether you liked a speaker or not, but did the speaker contribute to the day's learning objectives and was the lesson effective in terms of the course and institutional objectives. The last responsibility is also linked to membership within a profession. As USAWC graduates, our students accept the responsibility to continue to study and learn about the profession. Throughout the remainder of their careers, our students must continue to study and practice refining their ability to think and act at the strategic level.

In a typical military academic year at the USAWC, faculty members cannot teach strategy to everyone; nor can most students learn to be strategists proper. Not only is strategy difficult, but the limitations on resources and other legitimate demands on student time are preclusive. What faculty and students can do together is create an environment in which all can learn at differing levels the science and art of strategy. Faculty members can build on what has gone before and lay a foundation of knowledge and habits of thought for continued learning and practice that will serve graduates in their future roles. Teachers can also establish a common understanding of theory, process, and strategic thought that facilitates graduates working collaboratively and effectively in the national security arena. Some students may blossom during the year into recognized strategists; others may blossom later with additional study and experience. What is clear is that students who accept responsibility and apply themselves in learning at this level of education are better prepared to apply the strategic art in confronting the future challenges they face personally, and the challenges the professional faces collectively. Knowing this, both faculty and students as members of the profession must seek to create the best educational experience for all.

ENDNOTES - CHAPTER 7

1. Sun Tzu, *The Art of War*, Samuel B. Griffith, trans., New York: Oxford University Press, 1963, pp. 66, 91-92, 93.

2. Carl von Clausewitz, *On War*, Michael Howard and Peter Paret, eds. and trans., Princeton, NJ: Princeton University Press, 1976, pp. 140-141, 146-50, 578.

3. *Ibid.*, p. 150.

4. Colin S. Gray, *Modern Strategy*, Oxford, UK: Oxford University Press, 1999, pp. 1-3, 52-54, 79-83, 120-121.

5. *Curriculum Catalogue*, Carlisle, PA: U.S. Army War College, 2009, p. 4.

6. Richard A. Chilcoat, *Strategic Art: The New Discipline for 21st Century Leaders*, Carlisle, PA: Strategic Studies Institute, U.S. Army War College, 1995, p. 3. (See n.5, p. 23.)

7. *Ibid.*, p. 8.

8. Clausewitz, p. 141.

9. Charles Beitz' contributions were made by his influence as a department chair on his faculty and his own studies and lectures at the USAWC. He gets no mention in Harry P. Ball's *Of Responsible Command: A History of the U.S. Army War College,* Carlisle, PA: Alumni Association of the U.S. Army War College, 1984, Rev. 1994. See Harry G. Summers, Jr., *"On Strategy: The Vietnam War in Context,* Carlisle, PA: Strategic Studies Institute, U.S. Army War College, 1981. See *Of Responsible Command,* p. 461, for a brief description of the fortuitous coincidence of Michael Howard's and Peter Paret's modern translation of *On War* and Summer's monograph.

10. Arthur F. Lykke, Jr., "Toward an Understanding of Military Strategy," chap. in *Military Strategy: Theory and Application,* Carlisle, PA: Department of National Security and Strategy, U.S. Army War College, 1989, pp. 3-8.

11. For a sampling of this rich faculty literature, see any version of the *U.S. Army War College Guide to National Security Policy and Strategy* or the earlier *Military Strategy: Theory and Application,* both published by the Department of National Security and Strategy, U.S. Army War College, pp. 3-8. Query the SSI website for the excellent monographs by Jablonsky and Metz. For an excellent example of Dorff's thinking, see Robert H. Dorff, "Strategy, Grand Strategy, and the Search for Security," in *The Search for Security: A U.S. Grand Strategy for the Twenty-First Century,* Max G. Manwaring, Edwin G. Corr, and Robin H. Dorff, eds., Westport, CT: Praeger, 2003.

12. For the most complete perspective in this regard, see Harry R. Yarger, *Strategy and the National Security Professional: Strategic Thinking and Strategy Formulation in the 21st Century,* Westport, CT: Praeger Security International, 2008.

13. Sun Tzu, *The Art of War;* Clausewitz, *On War;* Gray, *Modern Strategy.* (See notes 1, 2, and 4.)

14. Clausewitz, pp. 100-102, 147-149.

15. *Ibid.*, p. 585.

16. *Ibid.*, p. 140.

17. *Ibid.*, p. 147.

18. William Duggan, *Strategic Intuition: The Creative Spark in Human Achievement*, New York: Columbia University Press, 2007.

19. *Ibid.*

20. William Duggan, available from *columbiapress.typepad.com/strategic_intuition/2007/08/from-william-du.html*.

21. *Ibid.*

22. Hilary Austen Johnson, "Artistry for the Strategist," *Journal of Business Strategy*, Vol. 28, No. 4, 2007, pp. 13-21.

23. *Ibid.*

24. *Ibid.*

25. *Ibid.*

26. *Ibid.*

27. Clausewitz, pp. 595-596. The center of gravity quote about faculty is common among commandants and deans at USAWC.

28. Major-General J. F. C. Fuller, from his first lecture at Camberley Staff College, 1923, quoted in *Memoirs of an Unconventional Soldier*, London, UK: Nicholson and Watson, 1936.

CHAPTER 8

THE TEACHING OF STRATEGY;
LYKKE'S BALANCE, SCHELLING'S EXPLOITATION,
AND A COMMUNITY OF PRACTICE IN
STRATEGIC THINKING

Thomaz Guedes da Costa

In the modern use of force, a squad leader stopping a vehicle at a checkpoint or entering a village in Colombia, Haiti, or Iraq must be cognizant of the scope of possible engagements he will experience at that moment and the consequences of the decisions and actions he will undertake. On the scene, his combat training must go hand-in-hand with his ability to negotiate and discern what is going on and what may come thereafter. The immediate objective of the mission may be clear to him, his team, and his superiors. But the nature of the mission may be volatile and mutate as the operation progresses. For the soldier and what he represents, his success in building a course of decisions and implementing actions will open new possibilities. Whether the soldier uses force or not, he has the power to either improve the political situation or risk damaging it, just like his generals, if on a different scale. How does the soldier know what to do? How did the warrior learn in order to be militarily and politically successful as he faces a change in the environment of warfare?

This chapter explores some essential questions regarding the teaching of strategy. It is partly a result of the dialogue of the Teaching Strategy Group (TSG).[1] This dialogue includes an evaluation of interaction among contemplation, classroom activities, curriculum design and development, and policy and political issues in defining educational purposes and professional competencies. This author humbly attempts to navigate the inception, execution, and evaluation of individual adult instruction for military and civilian officials as well as nongovernmental workers in the art, science, and craft of strategy in security affairs. In addition, the following analysis also challenges some assumptions of what strategy is and how social knowledge about strategy is structured in the official national security environment in the United States.[2] It also aims to raise questions and suggest answers, about the value of educational efforts in the discovery, learning, and application of knowledge to practical, nonstructured problem solving in a framework where the idea of strategy is central. The discussion examines the hypothesis that a particular preference about the *nature of strategy* will impact on both the environment that shapes a community of practice or of knowledge[3] and the preferred curricula comprising *what needs to be taught*.[4] This chapter explores also the role of the Skelton Report of 1989[5] in validating a particular understanding of what strategy is. Finally, it shows a sampling of institutional experiences and decisions that illustrate the interplay between a dominant *concept of strategy* and the definition of competencies for strategists within educational services.

THE NATURE OF STRATEGY AND THE NATURE OF THE STRATEGIST

Do We Need to Study Strategy?

Bernard Brodie was intrigued by the influence of education on the rise of victorious generals and statesmen. Consequently, he wondered how those who reached the top learned *strategy*? In a closing chapter of *War and Politics*, the author did not have an answer, but argued that until late in the 19th century, most strategists with experience in conducting warfare were never really formally educated for the type of thinking and the magnitude of the decisions they would have to make. Brodie's attention was not on formative schooling for military officers or politicians. His concern was the problem that would be settled at higher level, at the level of national politics or leading armies. Brodie posed that a "commander received little benefit from unstructured experience or from word-of-mouth instruction,"[6] while innate talent and foresight afforded some generals the opportunity to fight successfully. Heads of state also had little formal education about strategy and war. Moreover, as earlier thinkers such as Niccolò Machiavelli and Maurice de Saxe, or later Clausewitz, compiled and coded methods for militaries and war, Bernard Brodie questioned who would take the time to read those works or to reflect on the problems of war or international politics, and who would build intellectual constructs upon the experiences and insights of others into the decisionmaking processes in order to deal with new problems.[7] Would strategists-in-the-making have the time or willingness for study and contemplation before application? By the late 19th century, either under a liberal or a barracks education, many individuals would have the opportunity to refine their intuition, tacit knowledge, and innate abilities regarding the highest level of the use of military force, from both the multitude of recorded observations and organized instruction in military schools. Progress continued through the 20th century. But a question still lingers: Does formal education in strategy matter?

If one must consider the role of structured education in relation to strategy, the actuality of Brodie's earlier work shows the same questions that still puzzle some at the dawn of this century as war manifests itself in different characteristics and forms. Generally speaking, one can tease out many of the challenges that go along with teaching *about* politics, strategy, military history, diplomatic history, national security, economics, and other subjects, if one should use the Socratic Method, the case study technique, or simulation exercises. One can also explore, even under the perspective of individuals developing a career-long consciousness for continued education in strategy, when, what, and how he or she can be immersed in a structured education activity and what the core of such a program is for providing competencies to the strategist. Further, one can even dwell on the relationship between practical experience and academic learning as if they were two separate entities. Beyond Brodie's fundamental concern about learning strategy, issues abound, especially about the proper depth and scope of curricula, the sufficiency of core subjects, the selection of instructional methodologies, and the measurement of learning outcomes.

This problem of what and how to teach about strategy permeates the education of those in the military, the diplomatic service, or in business. Even if innate abilities and geniality may flourish, this author accepts the assumption that the smart individual learns from experience, and the wise individual from all accessible experiences, including those of others. Further, this author also would expect that institutional clients, stakeholders, and faculty at professional schools would devote close attention to their own input and respective impact on furthering education. The need for an introspective evaluation of the teaching of strategy is both academic and political because the aim (and the hope) is to achieve increased knowledge and value in the outcomes of decisionmaking. As Richard K. Betts indicates when discussing the reality and illusion of strategy, successful strategists understand that counterintuitive reasoning and the elusiveness of strategy demand intrepid mental and physical movements, the embrace of luck, and confident judgment, especially when confronted with securing absolute objectives when the use of some coercive force seems to be the only option available *a priori*.[8] If rigorous judgment of the environment and of interactions is advisable in overcoming or supplementing imperfect information or limited inductive or deductive reasoning, there is room for structured instruction, at least in order to induce the ability of the strategist to learn more and increase his chances of success.

The Strategist, not as Analyst, Planner, or Manager, but as Synthesizer.

It is a common tenet among thinkers that strategy has many levels of engagement.[9] Both in politics and business, for the two professional areas normally referred to as the pure *loci* of strategy, the actor is engaged in multilevel thinking. Overall attention must turn first to the interactions among actors, either in a cooperative or a competitive mode, where each participant is simultaneously making decisions and projecting movements. But in the background, actors can affect the environment, as they seek to shape intrinsic features or conditions, by changing norms or rules, or by changing structural variables defining the nature of the interactions. At both levels, a basic question to assert the value of structured education is *what is the core of the strategist's function?* Is the strategist a professional who *provides understanding* about a situation related to one's own critical objectives? Or beyond this, is the strategist *a synthesizer?* The synthesizer would be one who reaps new insights and conceives promising, viable solutions to problems. His aim would be to address not only the objective parameters, but also the subjective ones related to the situation at hand, while also enabling the shaping of the environment's structure, combining multidimensional thinking, as Ross Harrison describes in this book.[10]

Bernard Brodie's evaluation in *War and Politics* is enlightening because it pins down how the scientific approach of system analysis, in many variations, became a dominant pillar of the learning community in defense in the United States, seeking efficiency in capabilities and maximizing the utility of resources and engagements in the management and politics of the use of force. A link can be made between Brodie's proposition on system analysis and Richard K. Betts's warning that there is such a thing as the artless use of force that would reduce the value of serious strategizing.[11] Further, if one actor perceives a clear superiority over others, then the actions based on the overwhelming power of one's capabilities would "favor fully prepared offensives and frontal attacks."[12] In this case, why think beyond kinetics and one's own ability to dominate rivals?

In Betts's opinion, as one actor achieves overwhelming power, especially in conducting conventional war, the strength of a strategy matters little, since attrition campaigns may suffice if the objectives are seen as vital and costs are sustainable. Betts moves on to suggest that this reality of a powerful country, and its acceptance of costs and burdens, explains the U.S. preference in directly seeking decisive engagements, while expecting to outlast others in attrition matches. Therefore, if one were to fuse the argument of system analysis dominance with the reality of American military power, one would not expect to find creative maneuvers leading strategic thinking in the U.S. establishment, but would rather see an ever-increasing demand for improving efficiencies through self-examination.

Published at the height of the Cold War, Brodie's writing refers also to the split function of the strategist in the modern era of national security and defense. He argued about the complexity of intervening variables molding the selection of options in decisionmaking. The role of formal education on war and the shaping of a world view from experience do not stand alone (in terms of the individual's cognition and emotions) in defining decisionmaking. Rather, organizational politics and bureaucratic practices and preferences, the theoretical contributions made by academics, the observance of career demands for individuals in military, diplomatic, and defense affairs, technological innovation, and the political struggle of who controls strategy in modern warfare combine to end strategic practice by a single commander.[13] Brodie pointed out that after World War II, strategic thinkers, planners, and decisionmakers divided into different breeds. Although intertwined, each group has its own path to organizational posts, distinct responsibilities, and specific opportunities regarding the practice of strategy. A greater demand for specialization shaped, albeit roughly in most defense organizations, a new set of values and preferences for the sufficiency of planning. While the competencies for planners have become understood, for the strategist they are more diffuse and uncertain. Although Brodie is a forceful proponent of the similarities of strategic thinking and other applied sciences, using a scientific approach as an effort to discover abductive[14] justification and to prescribe action, he distances himself from identifying strategy as a pure science. For him, the dominance of the new scientific strategists as system analysis overlapped with operational analysis and established the methodology for the creation, organization, and socialization of knowledge; for directing plans, processes, and criteria for action; and for evaluating U.S. national security formulation and implementation.[15]

Another glimpse into the nature of a strategist is highlighted by the care that B. H. Liddell Hart took in distinguishing *between the concept of strategy and the concept of what a strategist has to think about.*[16] In formatting his understanding of strategy, Liddell Hart espoused that integral relations are fused into balancing ends and means[17] with the effort of thinking about the other actors in that circumstance of the interaction.[18] Hence, Liddell Hart provides an integral description of the strategist's function through his purpose or aim, and not through the pure practice of strategy in itself:

Let us assume that a strategist is empowered to seek a military decision. His responsibility is to seek it under the most advantageous circumstances in order to produce the most profitable result. Hence *his true aim is not so much to seek battle as to seek a strategic situation* [I suppose he has a license for tautology] *so advantageous that if it does not of itself produce the decision, its continuation by a battle is sure to achieve this.* In other words, dislocation is the aim of strategy; its sequel may be either the enemy's dissolution or his easier disruption in battle."[19]

This chain of reasoning, weaving together the propositions of Liddell Hart, Brodie, and Betts, implies that shaping the formal and informal education of a strategist depends greatly and equally on what the purpose of the strategist is and on exposing the learner more to thinking about interactions in a particular context than to contemplating the overwhelming command of his own situation to increase efficiencies of possible practices, including those that may end in defeat. However, the function of the strategist is not an academic matter, but rather, it is an institutional decision.

The evaluation of the teaching of strategy must be developed simultaneously with the implementation of the practice. Setting aside the artless use of force, there may be other key reasons why the teaching of strategy may be irrelevant. In a competitive, complex social environment, the evaluation of educating certain individuals may be impractical to achieve. In Darwinian environments — such as those of games, markets, politics, or wars — the nature of the structure logically determines that some athletes, generals, statesmen, or businessmen will rise within the ranks and eventually will prevail successfully over others in conflicts within the same institution, and will lead it to interactions in the environment. In modern times, most of these individuals will have some sort of formal structured education or coaching, and they can take advantage of learned knowledge and skills for reflecting upon the experiences or problem solving of others. How does one measure the effectiveness of the educational intervention in the career trajectory of those that will command or decide? Can one design an evaluation that provides assurances that a particular curriculum provides significant gains over others for the high level thinker demanded for the conduct of national security affairs or corporate affairs? At the same time, how is it that this same educational evaluation refutes the comparative validity of one instructional technique against another? Does the teaching of strategy indeed matter?

Does the Definition of Strategy Matter? Do Definitions Reveal the Core of Education for the Strategist?

One finds varying propositions for defining strategy.[20] Generally speaking, in the Department of Defense (DoD) community of knowledge, the definition of strategy follows less its etymology (the art of the *strategos*) and more its functionality. And strategy refers usually to its application: "...strategy is the art of applying power to achieve objectives, within the limits imposed by policy."[21] As this analysis explores the relationship between the connotation of the term and knowledge, it aggregates the understanding of strategy in two sets, limiting the scope of conclusions in proposing to connect discourse, on the one hand, with implications of accepting different ontological frameworks in the teaching environment.

One approach is represented by what can be labeled Lykke's Balance, after the enduring influence of leading U.S. Army War College thinker and teacher, Arthur F. Lykke, Jr. Based on preceding classical thinkers, Lykke's approach gained the status of an authoritative theory precisely because of its widespread acceptance in shaping a thinking framework that Dr. Gabriel Marcella calls "the American way of strategizing."[22] Along with the dominance of system analysis preferences, a community of intellectual reflection sprouted up around the approach and was adopted almost as official preference by the core community of knowledge (and practice) of the DoD. In other words, while the theory in itself was not innovative at birth, it became powerful through institutional sanctioning.[23] It is elegantly stated as an equation of balancing ends, ways and means. This author attributes the second vision or paradigm, to Thomas Schelling, who argued that the core of strategy is based on one's expectation about the behavior and expectations of others, and how to exploit it for one's own advantages.[24] This "Schelling Exploitation" is also at the core of the strategy that is used in the world of markets and businesses.

Before expanding considerations of these two approaches and the consequences of their adaptation to education, this analysis must acknowledge Anatol Rapoport's suggestion that the medium or nature of social interaction in which strategic thinking is applied has consequences in defining the core, or nature, of strategy itself.

Especially in education, the concern regarding the definition of strategy matters. The definition establishes expectations of what the thinker (the strategist) produces. As bureaucracies, armies, businesses, and individuals strive to provide educational services, institutionally demanded competencies and career profiles command and shape curricula and learning objectives. In totality, these institutional elements set the expectations of those professional competencies needed for individuals entering career paths and activities in different steps of the same institutions. As definitions are adopted, each preference sets the discourse and induces analytical boundaries for solving problems for a learning community or a community of knowledge as already mentioned. The assumptions about this community, in its potentials and limitations, is the fundamental puzzle that provokes one to think about the quality of learning and that of defense and national security decisions.

A CAVEAT: DIFFERENT DOMAINS OF STRATEGIC INTERACTION COULD MAKE A DIFFERENCE IN HOW ONE THINKS ABOUT STRATEGY

It seems that Anatol Rapoport was so concerned about how the environment would shape interactions that he paid little attention to valuing or using a rigorous concept of the term "strategy" itself. In his analysis of conflict, Rapoport provides an elaborate discussion arguing that these social interactions reveal themselves in three basic domains or structures.[25]

The first strategic environment consists of *debates*. In his view, debates are clashes of convictions — the essence of the conflict reveals itself in the effort of each debater to convince others of one's own superior or preferred proposals, beliefs, desires, or outlooks. Rapoport emphasized that in a debate, one can employ equivalent techniques and stratagems typical of fights and games. Debates display the exchanges of logic, promises and threats, justifications, and fundamentals of truth and beliefs, in addition to portraying the clashes of will, objectives, convictions, and preferences. From the output of rhetorical clashes, the final outcome connects one losing opponent to a *real* shift of convictions to the other's prevailing set of values. In other words, there is an intrinsic shift of value (or utility) as some players acquire or accept the new understanding of the truth (or what players or third parties think the truth is) and a movement towards the player that *wins*. New iterations of the debate will occur, and debates may have a permanent presence, as one observes is the case in politics and markets. Negotiation is a hallmark of debates.

Second, Rapoport reminds us that *games* are based on the agreement of opponents "to strive for incompatible goals within the constraints of certain rules."[26] The incompatibility of goals and the agreed rules or regulations reduce, but do not eliminate creativity of movement and courses of actions to overcome the will and capabilities of others to pursue the same exclusive goals. For Rapoport, opponents in a game are in a mirror image situation, striving to outwit or outperform each other. The acceptance of the structure and rules of the games limit the ability of the strategist to significantly shape the environment in which new iterations will take place. But there are many possibilities that affect the balance of will, perceptions, and attitudes, from past to future iterations.

As for *fights*, Rapoport claims that the essence of this type of interaction is for one to harm and debilitate the opponent, who needs to either disappear (figuratively) or be "cut down in size or importance" in a struggle.[27] The fight seems to be a perfect analogy for the art of the general, as long as the effort to outwit or outperform is not assumed to be based solely on power. In this case, the law of pure attrition may rule. Further, the physical clash may not take place if a player's (or a general's) own evaluation convinces him that defeat at the onset of the fight is inevitable, and thus the player bows out. In modern political conflicts interactions are multiple, and national strategy does not reduce itself to the essence of a fight according to certain goals, preferences, or capabilities.

Rapoport strongly proposes that one must seek and understand "different kinds of intellectual tools for the analysis of conflict situations."[28] Further, he warns that "no single framework of thought is adequate for dealing with such a complex class of phenomena as human conflict."[29] Rapoport was not concerned with strategy itself, but with the structures of conflict.[30] Nevertheless, his view of strategy was essentially operational, pointing out that strategy is a course of action, an argument about a general methodology of thinking that may be integrated into debates, games, and fights. Perhaps that was an intuitive forewarning about how colloquial the meaning of the word would become in many languages, how analogies of "how to do" as *schema* for information processing would flow from one dimension to the other, making strategic thinking either pervasive or useless.

BALANCING ENDS, WAYS, AND MEANS

In the U.S. Government, the dominant operational definition of strategy is summarized by Lykke's Balance. While Harry Yarger agrees with Gregory Foster that "There is little evidence that collectively as a nation there is any agreement on just what constitutes a theory of strategy,"[31] Lykke's proposal and eventual summary "is the centerpiece of this theory."[32] This conceptual approach of Lykke has been socialized in the sense that strategy is an equation that combines a balance of ends, ways, and means. In the same U.S. Army War College publication, his co-author, David Jablonsky, affirms that the "ends-ways-means paradigm also provides a structure at all levels of strategy to avoid confusing the scientific product with the scientific method."[33] While Lykke's own concerns were clearly addressed to the military strategy level, he argued that his approach is applicable at the national (grand) strategy and operational levels.[34] The balancing equation or theory has settled as the primer for organizing the efforts for political and military entities to seek respective objectives of policy. It became a solid fundamental for authoritative demands and educational frameworks, guidance and manuals.[35] If that is the doctrine and the intellectual preference of an institution, how does one know there is a need to shift paradigms?

This analysis does not study the relationship between Lykke's ontology and the institutional application of this methodology in doctrine and in problem solving in the schoolhouses of the DoD.[36] What educators and stakeholders concerned with the teaching of strategy must ask is what bias such a wide ranging analytical and institutional preference may have caused in terms of generating a closed community of educators and policymakers observing a common conceptual beacon and a shared quasi doctrine both in definition and application. Thomas Kuhn asked: "What do its members share that accounts for the relative fullness of their professional communication and the relative unanimity of their professional judgment?"[37] If the prime goal of Lykke's Balance theory is to *skillfully* balance the equation, would it encourage greater

217

attention on efficiency, to take advantage of means? Or would it direct efforts more to survey the environment and players, to adjust ways? To what extent does the assessment of others and the environment dominate the reasoning? If resources are abundant, would it induce the artless use of force while preferring system analysis practices for defining strategy, as Brodie indicated?

Furthermore, the powerful element that one ought to consider in relation to Lykke's Balance is its potential relationship to the establishment of a schema, or mind-set, for information seeking and processing at the individual cognitive level.[38] Such *schemata*, the balance of ends, ways, and means, permit shortcuts for filtering and interpreting new information and provide a quick reference not only to the individual but also for discourse, information sharing and methodology in any and all situations.[39] Lykke's Balance can function as a practical *schema* for problem solving, including at any level or type of pursuit of objectives, as Robert Dorff affirms in his primer on strategy development.[40] Thus, the concept of strategy may lose value because it could become merely a process for solving any problem, thus becoming a plan, a program, or even a recipe, as clearly indicated by J. Boone Bartholomees, Jr.: "In my own view, strategy is simply a problem solving process. It is a common and logical way to approach any problem – military, national security, personal, business, or any other category one might determine."[41] Therefore, what is the function of the strategist, returning to Brodie's and Liddell Hart's concerns? Or perhaps one could further ask that if all individuals are strategists (per Bartholomees' argument), what is there to teach? In this case, what is the core focus of a curriculum for military and national security officers and what distinguishes it from problem solving at, say, a medical school? This may explain why many curricula seem to hold the teaching of strategy equivalent to the teaching of the U.S. national security process.

FOCUS ON EXPLOITING THE SITUATION

Another approach to the essence of strategy lies in the perspective that what dominates its core is the interaction among actors, where the value of a decision for one depends upon the decisions of others, and strategy addresses how to influence decisions. This author suggests that Thomas Schelling's Exploitation represents one of the best representations of this approach, especially for the wide acceptance of a unifying framework in social science for understanding conflict with direct application to the policy sphere, with his inestimable influence in shaping strategic thinking in decisionmaking during the nuclear age.[47] If Lykke's Balance dominates the official U.S. national security and defense establishment, Schelling's approach concerning the environment and the behavior of others finds wide acceptance in market and business communities.[43] It clearly identifies with Rapoport's propositions regarding debates and games. Further, it addresses fighting in the essence of the *strategos*. Schelling's core arguments address the *problématique* of many military commanders, diplomatic officials, and strategic thinkers who pay attention to international politics, military affairs, and national security. For Schelling, strategy:

> is not concerned with the efficient *application* of force but with the *exploitation of potential force*. It is concerned not just with enemies who dislike each other but with partners who distrust or disagree

with each other. It is concerned not just with the
division of gains and losses between two claimants
but with the possibility that particular outcomes are
worse (better) for both claimants than certain other
outcomes . . . to study the strategy of conflict is to take
the view that most conflict situations are essentially
bargaining situations.[44]

In terms of historical perspective, this approach
certainly prevailed in circles of U.S. strategic thinkers
during a time when concerns over the actual use of
nuclear weapons reduced the viability of nuclear
power states using direct employment *in lieu* of
indirect strategy. Deterrence, compulsion, and coercive diplomacy came forward as strategies by the
limitations in the artless use of force, especially by the
uncertainty of escalation and the actual use of atomic
weapons.[45] Direct military confrontations promised
little under the edge of the overwhelming military
power umbrella, and the value of force moved into
the equation of how to exploit perceptions of will and
capabilities of the potential use of force. What makes
Schelling's thinking powerful not only presents a core
for the strategic problem but also permits the deduction
of a series of propositions in terms of realistic bargaining
tactics that any operator can use in noncooperative
interactions.[46] From nuclear weapons concerns to
irregular warfare, Schelling's synthesis is not only
sympathetic, but also corroborates the modernity of T.
E. Lawrence's *Pillars*, chapter XXXIII.[47] The essence of
thinking strategy in the post-Cold War conflicts is the
maneuver, the projections of movements in uncertain
future to secure new positions, resources, will, and
opportunities to reconstruct one's own posture facing
the environment. This is the dominant challenge for
heads of state, generals, and junior officers out in the
field in the modern use of force.

This anticipation of a rival's, adversary's, or partner's response in an interaction demands that one seek information, design maneuvers, and project decisions, movements, and courses of actions for future moments. Context, like scenarios, is not structured to provide a hypothetical situation to show the player how one can best perform in that particular situation or contingency. Contexts are shaped, like creeks in a flood basin, following trees and branches of possibilities and integrating intervening variables and new input. Successful players accept complexity mindfully and understand that the higher the level of decisionmaking, the greater the ambiguity in understanding the environment. Awareness of the situation and an understanding of projected vulner-ability and/or advantages are paramount. As Liddell Hart denoted, the strategist's need to think in terms of future engagements and advantages is the central tenet of his function.

The adoption and evolution of this paradigm has advanced game theory and its many variations into what is also known as the most common structured strategic thinking, with applications in all domains of human interaction. But it goes beyond this. Its powerful descriptive structures of relationships reveal to the player the heuristic value in organizing information and projecting abductive knowledge, especially in the realm of an intelligence analysis of motivation and of the creation of opportunity. It parsimoniously assists in teaching about complex situations in international politics as well as characterizes everyday situations as found in the marketplace, in business, and in other environments that include bargaining exchanges and negotiations.[48] For that, one expects that curricula would integrate instrumental subjects such as cognitive psychology, behavioral economics, logic, conflict management, game theory, and cultural and organizational awareness with those of the national security process, history, political science and international relations.

In Schelling's Exploitation, the nexus of strategy is centered on the dialectics of the will and ability to use force to resolve conflicts, not necessarily to seek an actual, decisive clash, as points out another classic strategic thinker, André Beaufre.[49] This approach, also preferred by thinkers such as Edward Luttwak, additionally incorporates at the forefront of reasoning the idea that decisionmaking interacts with the possible or actual use of force. Schelling's Exploitation seems to retain great faithfulness to the etymology of the concept of the *strategos*, while expanding the application beyond the scope of military affairs and retaining the core of politics – power play. This connotation of strategy also induces one to frame a particular conflict, struggle or dispute within an attitude of much more keen environmental awareness in which the interaction takes place. While it respects different levels of thinking and applicability (as Luttwak described), it highlights the multidimensional possibilities of engagement and maneuvering with direct and indirect approaches, as Beaufre indicated, in order to prevail.

WHY DOES THE ETYMOLOGIC CORE OF STRATEGY AFFECT COMPETENCIES AND CURRICULA FOR STRATEGISTS?

To advance an answer to the question, this section identified two of the dominant connotations of strategy in the intellectual and policy environments in the United States. The author does not recognize or claim that these two selections are the best or that they sufficiently represent the scope of all definitions of strategy, nor do they reflect a superior practical value or theoretical robustness over others. But these two approaches seem to embody different paradigms, in a Kuhnian sense. And because of that, the inter-paradigm dialogue and potential convergences between the two are not only possible, but are also promising in terms of exploring an improvement in the education of strategists. This seems to be especially relevant when such curricular integrations reveal to students and professionals a new menu of possible critical thinking on courses of actions when the responsibility refers to non-negotiable objectives of national security. In other words, if the objective of destroying or neutralizing al-Qaeda is non-negotiable, how does one develop strategies if the means are scarce and the ways are not creative enough to potentially recombine the same means?

The Skelton Report.

Another vein of reasoning to be explored in curricular teaching on strategy is to reflect upon the extent to which the Skelton Panel helped to enfold the education of a significant part of the DoD leadership by leaning its authoritative weight on one epistemological preference. In 1989, the House Armed Services Committee, led by Representative Ike Skelton, convened a panel on military education. As it sought information and opinions on the roles and expectations of professional military education (PME) in the post-Vietnam era, in the wake of the Goldwater-Nichols Act and reorganization of the DoD, this panel posed some general questions on the education of strategists: "How important is education? What type of education is relevant? What are the roles of PME schools as compared to other institutions? What type of faculty is needed?"[50] In their conclusion, the panel acknowledged that a successful strategist displays an innate talent and capacity for self-learning from experiences. Nevertheless, it called for the necessity of a structured education, including on strategy, and provided many proposals that would influence the National Defense University.[51] But the essence of the recommendations for a curriculum converged on the issues of service and joint command, field experience, and a robust knowledge of the national security process. It observed the challenge of defining the breadth and scope of curricula, types of educational practices, and student evaluation techniques.

In the flow of this analysis, one must turn one's attention to evaluating the extent to which the panel's assumption framed a preference about "what is strategy," and therefore, by its vested authority, may have conditioned the nature of strategic thinking in national security thereafter.[52] This may sound like a long thread, running from a congressional hearing to the classroom. But it may not be — it demands further structured research and evaluation of sets of homogenous sources. Nevertheless, the indication is that the Skelton Report's acceptance of Lykke's Balance, including through its citation of *Joint Chiefs of Staff Publication 1.02* as a base line for the *understanding* of the concept, probably reinforced the dominant view that may have solidified the institutional educational canons and doctrinal determinations.[53] The advancement of Lykke's Balance may have reduced the overall attention of educational leaders on other possibilities of thinking about strategy in the study of warfare, the national security process, and defense policy and management. The suggestion made by the Report may not have intended to be narrow and rigid, since it argued for the strategist to demonstrate ideal attributes such as analytical skills, pragmatism, innovation, and a broad education. The expectation was that, in the end, this individual would acquire the proficiency to afford the skillful balancing act of ends (objectives), ways, (courses of action), and means (resources) and would achieve in practice the tenets of system analysis promised, per Brodie's argument, and not the exploitation of situations, as Schelling would argue.[54]

Twenty years later, some indications suggest a state of flux. The main debate in strategic management assumes that more efficient governmental conduct, through much improved interagency coordination, ought to solve the strategic problem, whatever it is. As this line of reasoning seems to translate into the Army organization, the scope of work established for military staff duties and responsibilities fits well with Brodie's early alert, since the principal guidance is that officers will assist with strategic plans and policy.[55] This set of expectations moves closer to the argument that the strategist is an analyst for planning at the higher level, far from being the interactive player as Schelling's Exploitation would suggest, because he is inside the formulation process, but is not the decisionmaker. As for the Navy, on the other hand, the approach indicates a flexible adaptation (or recent evolution) as the environment becomes the focus of attention for the strategist in terms of education:" The task for strategists and planners in translating operational outcomes into enduring strategic results is never easy or straightforward. The Strategy and War Course at the

Naval War College examines how the overall strategic environment shapes operational choices and outcomes. In turn, the course also examines the strategic effects of operations, exploring how battlefield outcomes change the strategic environment."[56]

According to Ross Harrison's 3Ds proposal, at the lowest or grass root level, the individual is an operator who becomes a strategist when he or she needs to decide whether to shoot or negotiate in a small village somewhere. In the ever-increasing level of engagement, the strategist becomes a planner, being exposed to ever greater degree of complexities. Many issues here need further clarification, especially the function and validity of the education at the war college level, which was the focus of the Skelton Report. As views of the recommendations evolve, the issue of what is the nature of strategy that doctrine needs to observe also will evolve and, again, the Report will have its role in constructively shaping the teaching of strategy within the national security of the United States.

THE RELATIONS BETWEEN THE CHANGE OF DEFINITION AND THE CHANGE OF CURRICULUM

This author does not claim that there is a clear causal and unequivocal directional relationship between the change in the definition of strategy and adjustments in the core elements of a curriculum for the learning of strategy. Many intervening variables may be at play, including the rule of institutional guidance on competencies expected by stakeholders. Sometimes, the definition of strategy can have little impact if the establishment of competencies by institutional clients clearly enables and commands a curriculum design grounded in conceptual diversity and inter-paradigm dialogue. But in some occasions it may condition the scope of learning.

The Center for Hemispheric Defense Studies (CHDS), at the National Defense University, in Washington, DC, has had some experiences with the challenge of exposing students to models of strategic thinking and their practical implications, by addressing dominant approaches to the issues of "what is strategy" and the role of the strategist.[57] The year 2004 presented a convergence of challenges to CHDS' teaching effort. In that calendar year, CHDS taught the Defense Planning and Resource Management Course (DPRM), the Inter-Agency Coordination and Counterterrorism Course (ICCT), and the Curriculum Design and Instructional Methodologies (for National Security and Defense-CDIM). It also developed an interagency educational program to assist the Colombian government in implementing a model for improving national interagency coordination to support a new strategy of national territorial presence and control of key localities in that country. The Center also began to offer a structured workshop to respond to requests by Central American countries on improving national security processes.

Since its inception in 1997, CHDS's faculty has been comprised of an international team of professors who were educated and have had professional practice in the militaries, defense establishments, academic institutions, and governments of foreign countries. From its inception, the dominant view for the teaching of "what is strategy" was the integration of Lykke's Balance into curricula as the dominant framework of how to think strategically and how to frame the overall problems of national security, just like most other joint professional military education institutions in the United States.[58]

In the spring of that year, CHDS offered a unique course that challenged this notion.[59] Designed and led by the late Professor Ismael Idrobo, the Curriculum Design and Instructional Methodology (for security issues) was more than a teaching endeavor to serve students. It permitted the faculty to evaluate curricular development and integration, as well as to further improve the understanding of how to assist adult learners in thinking creatively about issues in defense and security. The course permitted some members of the faculty to explore contradictions between promoting high-level thinking in defense and security and the narrowing impact of the ends-ways-means constraints framework. In discussions and exercises, Lykke's framework would drive students to pay close attention to resource constraints as a point of departure, almost as a factoring policy principle, in most arguments for designing and implementing strategy at every level.[60] Inside the CHDS teaching environment, the late Felipe Rojas provided an early warning about this problem after studying the impact of the multiplier effect of network-centered thinking and the effort to fight terrorism. Rojas's arguments were integrating not only his past experience as a military officer in Peru, successfully fighting the *Sendero Luminoso* (Shining Path), but also projecting into the future the magnitude of the challenges of ideological extremism and conflict.[61] Some instruction provided to the students, particularly about defense policy and military capability, risked overshadowing and derailing the educational effort on the path to creativity in interactive modes, if it were not for the ability of instructors to fuse models and frameworks in thinking about strategy.[62] Time and time again, in nearly every course delivered at CHDS, as discussion turned to the issue of resource constraints, students seemed to move into a groupthink frame of mind, deadlocks, and paralysis by the salience of the idea that there were not enough resources to apply Lykke's approach properly.[63]

The development and execution of the curriculum design course showed, especially in its segment on terrorism studies, that Schelling's Exploitation would permeate discussion among participants as notions of network centric warfare, asymmetric warfare, root causes of terrorism, the experiences of counterinsurgencies, and the fusion of counterterrorism with insurgency, began to dominate dialogues both in seminar exchanges and at the water cooler.[64] The discussions were not simply referring to the clash of motivations, goals, capabilities, and opportunities, but of strategy and the battle of ideas as material resources becomes a secondary concern. Late in that same year, for all practical purposes of course instruction, the definition of strategy had shifted to approach the more robust Schelling's Exploitation, emphasizing the interactive dimension in strategic thinking, while preserving demands for efficiency and accountability expected in democratic societies.[65]

By 2005, the leading foundational course at CHDS had shifted from an emphasis on resource management in the defense sector. The defense management course (DPRM) was terminated and Strategy and Defense Policy Course (SDP) was started, aiming to provoke learners to advance abilities to think about issues of policy and strategy in the national security environment.[66] Since then, the Center has developed a National Security Process Workshop to assist national governments, especially from Central America, in the framing of policy and strategy problems in national security and defense. For these workshops, the curricular emphasis retained the Lykke's Balance approach to the thinking about national strategy. In this case, the intellectual preference of some professors for an interactive approach was restricted by an institutional preference, comfortable with the approach dominant within DoD's doctrine.

Educational institutions need to evaluate how these choices affect learning outcomes and competencies and, if possible, comparing different services both in terms of creating individual professional and intellectual performance and institutional satisfaction.

IN LIEU OF CONCLUSIONS, FURTHER STEPS ARE NEEDED

It is useful for institutions to establish operational definitions to implement doctrine effectively in their internal communication, education and training, and organizational socialization. However, the adoption of key concepts (and respective definitions) in an educational environment must take into account the heuristic value of such a particular preference, including evaluating how a choice may constrain intellectual and practical discoveries for problem solving. It was not the purpose of this chapter to compare competing definitions of strategy. Nor it was to evaluate the consequences of professional education for defense in the U.S. national security environment. In fact, such a study must take into account not just joint professional military education, but also civilian education because high policy is conducted mostly by political officials and not military commanders. The performance of countries and institutions in counterterrorism, stability operations such as those in Iraq, Afghanistan, Haiti, or Chad, and in adjusting organizationally to new manifestations of armed conflicts, demands attention of officials, officers, educators and other stakeholders.

As the Teaching Strategy Group's discussions move forward, the main purpose of this review was to explore the relationship between a choice in the ontological approach to the nature of strategy and the setting of some fundamental assumptions about what core knowledge matters to both individuals and institutions as they face professional challenges in the high level of reasoning in national security processes. The instructor must have a clear understanding about *what strategy is, what the strategist is,* and *how to provoke the student to think about how to conceive a design and carry it out in each structural dimension of interaction.* Then, further thinking and evaluation about *"how to teach"* can productively move ahead. If an instructor of strategy strives to evaluate the impact of his or her performance beyond the limits of the classroom or of established learning outcomes, one is caught in the conjectures regarding the effects of the teaching of strategy in itself and the constraints imposed by the institutional setting or the community of knowledge involved in the learning environment. The teaching space is defined by personal academic reflection, team discussion, the demanded competencies, and the framework of curricula. While there is the practice of building instruments to measure performance and satisfaction within the limits of instruction, there is a lack of institutional research that would embody further knowledge and bring about valid systemic evaluation of the practice. In this case, the evaluation even for this analysis suffered from the restriction of public access to curricula at most schools.[67] Therefore, more structural evaluation of the impact of teaching strategy is desirable.

Identifying a dominant paradigm in the discourse and practice of a community of knowledge in the U.S. national security environment allows stakeholders to ponder the intellectual possibilities and limits in meeting political responsibilities with analytical tools shaped by existing curricula. By the pervasiveness of Arthur Lykke's conception, this author does not doubt its powerful intellectual value. The importance of addressing this pillar of thinking about strategy is not based on the achievement and cognitive comfort that it has afforded to individuals and institutions. After all, the dominance of the United States in international security and in the preservation of its vital objectives has not come about by chance. The relevance lies, perhaps, in what this dominance defines in terms of bias, missed opportunities, and permitting full discovery for strategists to respond to new dimensions of conflict and the use of force. Therefore, a critique of the fundamentals of teaching strategy based solely on a single school of thought is prudent.

In reflecting on the subject of teaching strategy, one striking and lasting insight is the need to further seek out an improved understanding of the function of the modern strategist in wars such as those of Colombia, Haiti, or Afghanistan. There may be a relevant variety of career profiles. There is the planner and official whose main responsibility seems be the improvement of efficiency by focusing on the linkages of different levels of strategy design and implementation. There is the conceiver whose task, closely linked with action able intelligence, is to explore creativity and expectations in order to invite new ways of prevailing in interactions under time constraints and in unusual situations. And there is the officer in the field, as the squad leader, who needs all the intellectual and emotional support he can get, *a priori*, to navigate different dimensions of the strategy under the stress of the immediate success or failure of his decisions inherent to the complex nature of the conflict or the mission.[68] Together, through pride and professional accountability, those in institutional positions of responsibility and those in instructional environments must together explore how to improve, even if marginally, the teaching of strategy for the benefit of the new young "Aurens."

ENDNOTES - CHAPTER 8

1. For some of the ideas advanced by colleagues of the TSG, see Gabriel Marcella and Stephen O. Fought, "Teaching Strategy in the 21st Century," *Joint Force Quarterly*, Issue 52, 1st Quarter, 2009, pp. 56-60.

2. Obviously, the methodology of this chapter demands further consultation with primary sources, such as curricular guidance, syllabi, and evaluation tools that are not readily or openly available to the public at this time, to further validate its central hypothesis.

3. On the formation of communities of specialists, see Thomas S. Kuhn, *The Structure of Scientific Revolutions*, 3rd Ed., Chicago, IL: Chicago University Press, 1996, pp. 176-181.

4. Although I recognize the importance of advancing a deductive exercise on how each ontological view of strategy will suggest "what to teach," that effort is beyond the scope of this chapter.

5. U.S. Congress, House Committee on Armed Services, *Report on the Panel on Military Education*, 101st Cong., 1st Sess. 1989, Committee Print 4, p. 26. Hereafter *The Skelton Report*.

6. Bernard Brodie, *War and Politics* , New York: MacMillan, 1973, p. 435.

7. *Ibid.*

8. Richard K. Betts, "Is Strategy an Illusion?" *International Security*, Vol. 25, No. 2, Fall 2000, pp. 46-48.

9. For a solid analysis of strategy and its levels, see Edward N. Luttwak, *Strategy: The Logic of War and Peace,* Cambridge, MA, Harvard University Press, 1987. Marcella and Fought, pp. 57-58, also warns that the issue of "all levels of strategy" must be clearly discerned in thinking about teaching strategy. See also a wealth of applications in Miguel Alonso Baquer, *¿En qué consiste la estratégia?,* Madrid, Spain: Ministerio de Defensa, 2000.

10. Ross Harrison, *Teaching Strategy in 3D*, Paper presented at the XX Annual Strategy Conference, U.S. Army War College, Carlisle, PA, April 17, 2009.

11. Betts, p. 46.

12. Luttwak, p.15. Here Luttwak poses that the Allied offensives against a German army at the close of World War II shows that the logic of strategy was no longer important because the enemy's reaction could be ignored.

13. For instance, Brodie reflects on the intellectual developments at the eve of World War I: "Why had not the most obvious lessons of combat experience been absorbed by commanders who were to send great new armies into battle? Because for the most part experience not personal to themselves was not alive to these commanders, who were not students of history, even of military history, but who had absorbed an intense indoctrination laced through with religious fervor on the merits of offensive." Brodie, p. 457. For consubstantiation or bias and advice in the relationship between thinkers and decisionmakers in national defense, see also Richard K. Betts, *Soldiers, Statesmen, and Cold War Crises,* New York, Columbia University Press, 1991, pp. 183-212.

14. For an application of Charles Sanders Peirce's concept of abductive reasoning of "showing that something is plausibly true," see David T. Moore, *Critical Thinking and Intelligence Analysis,* Washington, DC, Joint Military Intelligence College, 2006, pp. 3-47.

15. Brodie, p. 453.

16. B. H. Liddell Hart, *Strategy,* 2nd Ed., 1967, reprint, New York, Meridian, 1991, pp. 319-333. This is the chapter entitled "The Theory of Strategy."

17. *Ibid.,* pp. 329, 332.

18. *Ibid.,* pp. 324, 327.

19. *Ibid.,* pp. 325.

20. For a survey of classical thinkers, see J. Boone Bartholomees, Jr., "A Survey of the Theory of Strategy" in J. Boone Bartholomees, Jr., ed., *U.S. Army College Guide to National Security Policy and Strategy,* 2nd Ed., revised and expanded, Carlisle, PA: Strategic Studies Institute, U.S. Army War College, 2006, pp. 79-106. See also Baquer, pp. 36-43.

21. Marcella and Fought, p. 57.

22. Dr. Marcella has often made this comment to the author in private conversations.

23. Michael J. Meese, "The National War College: Building Strategists for the Nation," in Jeff McCausland, ed., *Educating Leaders in an Age of Uncertainty, The Future of Military War Colleges, 2004-2005 Report to the Smith-Richardson Foundation,* Carlisle, PA: Dickinson College, pp. 272-273, available from *wwb.archive. org/web/20080306033855m_1/alpha.dickinson.edu/departments/ leadership/Smith%20Richardson/Smith%20Richardson%20 Report%20Page.htm.* Meese indicates the prevalence of Lykke's framework in the National War College, Washington, DC, as students are presented to the fundamentals of strategic logic:

235

"From the first day of the first course, students are introduced to the College's "ends-ways-means-constraints" framework that has now pervaded not only strategic thinking at the College, but much of the writing in U.S. national security documents" (p. 272). For a Joint Professional Education student's struggles and incisive reflections on the study of strategy and analytical framework for institutional use, see Gary B. James, *A Nonlinear Approach to Strategy Formulation*, Strategy Research Project, Carlisle, PA: U.S. Army War College, 2008.

24. Thomas C. Schelling, *The Strategy of Conflict*, Cambridge, MA: Harvard University Press, 1960, p. 5.

25. Anatol Rapoport, *Fights, Games, and Debates*, Ann Harbor, MI: University of Michigan Press, 1960.

26. *Ibid.* p. viii.

27. *Ibid.*, p. 9.

28. *Ibid.*, p. 12.

29. *Ibid.*, p. 359

30. It suffices to say that Rapoport presents his definition of strategy well in his study of conflict, when he points out that "... strategy in game theory is a technical term with a precise meaning. It means a complete program given by a player before the game begins, say to a referee, stating what he will do in every conceivable situation in which he may find himself in the course of the game..." *Ibid.*, p. 142.

31. Harry R. Yarger, "Toward a Theory of Strategy: Art Lykke and the Army War College Strategy Model," in J. Boone Bartholomees, Jr., ed., *U.S. Army War College Guide to National Security Policy and Strategy, Vol. I: Theory of War and Strategy*, 3rd Revised and Expanded Ed., Carlisle, PA: Department of National Security and Strategy, U.S. Army War College, 2008, p. 107.

32. *Ibid.*

33. David Jablonsky, "Why is Strategy Difficult?" in Bartholomees, ed., p. 143.

34. Arthur F. Lykke, "Toward an Understanding of Military Strategy," Joseph R. Cerami and James F. Holcomb, *U.S. Army War College Guide to Strategy*, Carlisle, PA: U.S. Army War College, 2001, pp. 179-180.

35. *The Skelton Report*, p. 26.

36. The author recognizes the need for such an evaluation, but most data, curricular structures (syllabi, etc.) are not publicly available.

37. Kuhn, pp. 182.

38. For the urgency of practioners in tackling problems and studying strategy, see Steve Fought, "The War College Experience." *Academic Exchange Quarterly*, June 24, 2004, pp. 1-2, as quoted in Gabriel Marcella, "The Strategy of Teaching Strategy in the 21st Century" *Of Interest*, Strategic Studies Institute, U.S. Army War College, November 8, 2007, p. 4.

39. Bradley S. Gibson, *The Convergence of Kuhn and Cognitive Psychology,* Paper presented at the Annual Meeting of the Rocky Mountain Psychological Association, Las Vegas, April 25-28, 1984.

40. Robert H. Dorff, "A Primer in Strategy Development," in Cerami and Holcomb, p. 11.

41. Bartholomees, Jr., "Survey of the Theory of Strategy" in Bartholomees, Jr., ed., p. 81.

42. Thomas C. Schelling, *The Strategy of Conflict*, Cambridge, MA: Harvard University Press, 1960, p. 3. At that footnote, Schelling warns that this concept "is not the military usage."

43. Some widely used primers are Henry Mintzberg and James Brian Quinn, *The Strategic Process, Concepts, Contexts, and Cases,* New York, Prentice Hall, 1991; and Avinash K. Dixit and Barry J. Nalebuff, *Thinking Strategically*, New York, W. W. Norton, 1991, pp. ix, 31. In fact Mintzberg and Quinn presented one of the most powerful blends of military and business thinking to synthesize strategy, pp. 3-22.

44. Schelling, p. 5.

45. Fred Kaplan, *The Wizards of Armageddon*, New York, Simon and Schuster, 1983; Lawrence Freedman, *The Evolution of Nuclear Strategy*, 2nd Ed., London, UK: Macmillan Press, 1989; McGeorge Bundy, *Danger and Survival: Choices About the Bomb in the First Fifty Years*, New York, Random House, 1988.

46. For a valuable summary of the development of Schelling's applicability, see Avisnash Dixit, "Thomas Schelling's Contributions to Game Theory," *Scandinavian Journal of Economics*, Vol. 108, Issue 2, 2006, pp. 213-229; or The Royal Swedish Academy of Sciences, *Robert Aumann's and Thomas Schelling's Contributions to Game Theory: Analyses of Conflict and Cooperation*, October 10, 2005, available from *nobelprize.org/nobel_prizes/economics/laureates/2005/ecoadv05.pdf*.

47. T. H. Lawrence, *The Seven Pillars of Wisdom, a Triumph*, New York, Garden City: Doubleday, Doran & Company, 1926, printing of 1936, pp. 188-196.

48. For instance, see Glenn H. Snyder and Paul Diesing, *Conflict Among Nations, Bargaining, Decision Making, and System Structure in International Crises*, Princeton, NJ: Princeton University Press, 1977. For market applications, see the updated work by Avinash K. Dixit and Barry J. Nalebuff, *The Art of Strategy, A Game Theorist's Guide to Success in Business & Life*, New York, W. W. Norton & Company, 2008.

49. André Beaufre, *Stratégie de L'Action*, (*A Strategy of Action*), Paris, France: A. Colin, p. 1966, p. 13.

50. *The Skelton Report*, p. 28.

51. John W. Yaeger. *Congressional Influence on National Defense University*, George Washington University, Washington, DC, 2005.

52. *The Skelton Report*, p. 25.

53. *Ibid.*, p. 26.

54. *The Joint Staff Officer's Guide, JFSC Pub 1,*Norfolk, VA: Joint Forces Staff College, 2000, pp. 2-3. As for demands on competencies to strategist and the core emphasis on national security processes, see Chairman of the Joint Chiefs of Staff Instruction, J-7, *Officer Professional Military Education Policy, OPMEP, CJCSI 1800.01,* Washington, DC: U.S. Joint Chiefs of Staff, December 22, 2005, pp. A-A-4 - A-A-8.

55. *Commissioned Officer Professional Development and Career Management, PAM 600-3,* Washington, DC, Headquarters Department of the Army, December 11, 2007, p. 263. It refers to Functional Area 59, individual work with plans and policy.

56. *Syllabus, Strategy and War,* U.S. Naval War College, College of Naval Command and Staff, Naval Staff College, November 2008-February 2009, p. 1. The author recognizes again that further evaluation of the linkages between definition and curricula needs to be conducted as access to sources becomes available. Information available from *www.nwc.navy.mil/academics/courses/ sp/documents/S&P%20CNW%20Syllabus.pdf.*

57. The Center for Hemispheric Defense Studies is one of the five regional centers of the DoD. Its mission is to educate and promote collaborative exchanges on matters of national security and defense, defense and national security processes, and international security. It serves mainly participants from the Americas in resident programs in Washington, DC, and academic events in the region. (Executive professional education 3-week courses are offered primarily to civilian government officials, military and police officers, members of legislatives and judicial branches, and to nongovernment professionals, university professors, academic researchers, journalists, members of political parties and other individuals directly involved in national security and defense matters). Course curricula are developed under norms of academic freedom.

58. Meese, pp. 272-273.

59. This awareness of change is, of course, very *post hoc.*

60. As full disclosure, one of the courses was Defense Economics.

61. The late Captain Felipe Rojas (Peruvian Navy) was a 2002 student at the Industrial College of the Armed Forces and later became a Visiting Scholar at CHDS. 2002-03.

62. I recognize the mentorship of Dr. Jaime García Covarrubias in the classics of military strategy, and Dr. Craig Deare in blending approaches to effectively instruct military strategy, military capabilities, and many approaches to strategic management of the DoD.

63. The author has personally observed this on many occasions.

64. This author was the leader of the Terrorism Seminar in the CDIM/2004. From that, he developed and directed the Interagency Coordination and Counterterrorism Course in 2004 and 2005.

65. For courses, CHDS has never adopted an authoritative definition of strategy.

66. The author led the design and development of the course and directed it since with the invaluable contribution of Manuel Lora.

67. As for full disclosure, such access may be requested.

68. *Post-script:* there are many descriptions of young American strategists, military and civilian, performing outstandingly in Afghanistan. For instance, see the accounts about Captain Michael Harrison in Greg Jaffe, "A Personal Touch in Taliban Fight, In the Afghan Mountains, a Company Commander Strives to Gain the Trust of Frustrated Villagers," *Washington Post*, June 22, 2009, available from *www.washingtonpost.com/wp-dyn/content/ article/2009/06/21/AR2009062102021.html?wpisrc=newsletter&wpisr c=newsletter&wpisrc=newsletter.*

CHAPTER 9

MAKING SENSE OF CHAOS:
TEACHING STRATEGY USING CASE STUDIES

Volker Franke

If the mind is to survive this constant battle with the
unexpected, two qualities are indispensable: first, an
intellect that even in this moment of intense darkness
retains some trace of the inner light that will lead it to
the truth, and second, the courage to go where that
faint light leads.

Carl von Clausewitz[1]

The secret of success in life is for a person to recognize
and be ready for an opportunity when it comes.

Benjamin Disraeli[2]

INTRODUCTION

In October 2008, the Army unveiled its new field
manual for stability operations,[3] thereby taking another
step beyond the military's traditional role of preparing
to "fight and win the nation's wars."[4] Indeed, the
U.S. military's mission today includes also winning
the peace under increasingly complex and uncertain
conditions. As Lieutenant General William B. Caldwell
IV states in his foreword to the new manual:

the lines separating war and peace, enemy and
friend, have blurred and no longer conform to
the clear delineations we once knew. At the same
time, emerging drivers of conflict and instability
are combining with rapid cultural, social, and
technological change to further complicate our

understanding of the global security environment. Military success alone will not be sufficient to prevail in this environment. To confront the challenges before us, we must strengthen the capacity of the other elements of national power, leveraging the full potential of our interagency partners."[5]

The Stability Operations manual reflects the doctrinal culmination of a national and defense policy evolution that began with the collapse of Communism and the fall of the Berlin Wall 2 decades ago.[6] The fact that peace and stability operations are now considered equally important as traditional combat missions represents a sea change for the U.S. military, both operationally as well as strategically.

Doctrine plays a special role in the U.S. Army—as evidenced by the fact that there exist more than 550 doctrinal manuals to date—providing fundamental values and principles, best practices and lessons learned to effectively fulfill its role in support and implementation of national objectives. Doctrine informs decisionmaking at the strategic, operational, and tactical levels. While the U.S. military excels in preparing its soldiers and officers for the operational demands and tactical requirements of a wide array of increasingly complex contingency missions, a number of observers have pointed to the need for teaching strategy more effectively as part of professional military education (PME).[7]

The purpose of this chapter is to examine how case study methodology can provide an effective vehicle for teaching strategy and strategic decisionmaking to military professionals. After an introduction to the concept of strategy, I examine in some detail the cognitive frames that inform strategic decisionmaking. Specifically, I discuss the importance of heuristic shortcuts as cognitive decision guides, and compare the rational actor decision model that has traditionally informed strategic decisionmaking in the military with a sense-making framework more suitable to complex strategic environments. Finally, I provide a brief introduction to the case study method and illustrate how case studies can be employed effectively to teach strategy in senior PME settings.

STRATEGY DEFINED

In military parlance, strategy refers to the maneuvering of troops into position before engaging the enemy. Carl von Clausewitz defined strategy as "the employment of the battle to gain the end of the war."[8] Strategy "must therefore give an aim to the whole military action, which must be in accordance with the object of the war; in other words, strategy forms the plan of the war, and to the said aim it links the series of acts which are to lead to the same, that is to say, it makes the plans for the separate campaigns, and regulates the combats to be fought in each."[9]

But what can a 19th-century Prussian general teach a 21st century military strategist? A great deal, in fact, given the similarity in the strategic challenges they both face: alliances are made, broken, and reconstituted at dizzying speed; when unprecedented events occur, experience does not necessarily indicate a course of action; and rules, principles, and "how-to" prescriptions may no longer apply. Today's strategic environment is characterized by threats that "are both diffuse and uncertain, where conflict is inherently unpredictable, and where our capability to defend and promote our national interests may be restricted by political, diplomatic, informational and economic constraints. In short, it is an environment marked by volatility, uncertainty, complexity, and ambiguity (VUCA)."[10] These, says Clausewitz, are the times in which true strategists thrive.

Examining wars from the time of the ancient Greeks through World War-II, British strategic thinker B. H. Liddell Hart criticized Clausewitz's conception of military strategy, since Clausewitz's famous dictum that war was the continuation of politics by other means made battle the only viable option for achieving strategic aims.[11] Instead, Liddell Hart favored German Field Marshal Helmuth von Moltke's definition of strategy, as "the practical adaptation of the means placed at a general's disposal to the attainment of the object in view," since that clearly connected strategy as a means serving political ends.[12] In all, Liddell Hart developed eight maxims of strategy, a number of which are still highly applicable to today's complex operational environment, including: constantly adjusting one's end to the available means with a clear sense of what is possible; recognizing and weighing the feasibility of alternative courses of action; being flexible and adapting to changing circumstances; and contemplating contingencies or next steps for successes as well as failures.[13]

Some of Liddell Hart's other strategic maxims, however, are less applicable. For instance, choosing the line of least expectation by figuring out the course of action that the opponent will view as least probable and acting in such a way as not to reveal one's objectives, may in fact undermine peacebuilding efforts. Contemporary contingency operations are often conducted through varying coalition constellations and require mutual trust and the support of the local population. Transparency, predictability, respect, and deference are cornerstones of effective peacebuilding and therefore must also be central components of strategic thinking and leadership.

In his 1994 book, *The Rise and Fall of Strategic Planning,* Henry Mintzberg distinguished four distinct, yet interconnected meanings of the term "strategy:"

1. Strategy is a plan, a "how," a means of getting from here to there.

2. Strategy is a pattern in actions over time.

3. Strategy is position, i.e., it reflects decisions to offer particular (military) services to meet particular policy demands.

4. Strategy is perspective, i.e., it offers vision and direction.

Mintzberg argued that strategy emerges over time as intentions collide with and accommodate changing realities. Thus, one typically starts with a perspective, concluding that it calls for a certain position, which in turn is to be achieved through a carefully crafted plan. The desired outcome and the strategy envisioned to achieve it is reflected in decision patterns and actions over time.[14]

Strategy, according to Yarger's recent *Little Book on Big Strategy,* "provides a coherent blueprint to bridge the gap between the realities of today and a desired future. It is the disciplined calculation of overarching objectives, concepts, and resources within acceptable bounds of risk to create more favorable future outcomes than might otherwise exist if left to chance or the hands of others."[15] Thus, strategy provides direction for courses of action intended to maximize positive and minimize negative outcomes.

Strategy, in other words, can be understood as a pattern of decisions that determines and reveals its objectives, purposes or goals, produces the principal policies and plans to achieve those goals while defining the position and the range of tasks and responsibilities of an organization, and specifies the contributions it intends to make toward accomplishing the overall mission objective(s).[16] Nickols defines strategy along four dimensions: perspective, position, plan, and pattern. It is,

> the bridge between policy or high-order goals on the one hand and tactics or concrete actions on the other. Strategy and tactics together straddle the gap between ends and means. In short, strategy is a term that refers to a complex web of thoughts, ideas, insights, experiences, goals, expertise, memories, perceptions, and expectations that provides general guidance for specific actions in pursuit of particular ends. Strategy is at once the course we chart, the journey we imagine and, at the same time, it is the course we steer, the trip we actually make. Even when we are embarking on a voyage of discovery, with no particular destination in mind, the voyage has a purpose, an outcome, an end to be kept in view.[17]

Strategy does not exist outside the ends sought. It serves as a general framework that provides guidance for actions to be taken and is itself in turn shaped by those actions. Thus, a clear understanding of the purpose and the ends pursued is a necessary precondition of any effective strategy. Strategy determines means; it is about the attainment of ends, not their specification. If strategy has any meaning, it is only in relation to the achievement of the end. Herein rests the great import of strategy and strategic thinking to the military. One of the defining features of the military profession is that its members ought to be above politics and seek to find the most effective military means to support the attainment of political objectives, without themselves taking a political stand.[18]

Conceived this way, strategy becomes part of a decision structure: First are the ends to be obtained. Second are the strategies for obtaining them, i.e., the ways in which resources will be allocated. Third are tactics, i.e., the ways in which allocated resources are actually used. Finally are the resources themselves, the means at our disposal. Thus, strategy and tactics bridge the gap between ends and means.[19] And effective strategy, Yarger conjectures, must be proactive. It is:

> fundamentally a choice; it reflects a preference for a future state or condition in the strategic environment. It assumes that, while the future cannot be predicted, the strategic environment can be studied and assessed. Trends, issues, opportunities, and threats can be identified with analysis, and influenced and shaped through what the state chooses to do or not do. Thus strategy seeks to influence and shape the future environment as opposed simply to reacting to it.[20]

The central challenge, Dorff conjectures is "adapting effectively to the new circumstances while simultaneously balancing against the lingering circumstances from the older system."[21]

THE STRATEGIC DECISION ENVIRONMENT

Decisionmakers today respond to a strategic environment quite different from that of the past. Characterized by VUCA, Yarger concludes "the strategic environment is always in a greater or lesser state of dynamic instability of 'chaos'."[22] As a result, decisions cut across a wide range of social, political, and cultural domestic and global issues and demand cognitive flexibility, adaptability and the ability to make decisions "on the fly." This means not only continuous learning on the part of individual military leaders, but also on the part of the U.S. military as an organization, reflected in the honing of strategic leadership and decisionmaking competencies.

STRATEGIC LEADERSHIP COMPETENCIES

In December 2001, the Chief of Staff of the Army tasked the U.S. Army War College (USAWC) to identify the strategic leader skill sets for officers required in the post-September 11, 2001 (9/11) environment.[23] The USAWC researchers set out to complete their task based on the assumption that future strategic leader capability, but not necessarily strategic leadership, was required at the brigade-level, since they believed senior officers (O-6) would need to think strategically, even if they are not in troop leading positions. Researcher Leonard Wong and his colleagues concluded that existing lists of strategic leadership skills were too comprehensive, since they required strategic leaders to "know and do just about everything."[24] Consequently, they suggested focusing on six meta-competencies with utility for developing strategic leader capability and facilitating self-assessment by officers of that capability:

1. Identity, including an understanding of one's self-concept as an officer in the Army (or any other branch) and of one's values as well as the extent to which those are compatible with the Army's.

2. Mental Agility, or adaptability, referring to the "ability to recognize changes in the environment; to determine what is new . . . and what must be learned to be effective."[25] Mental agility, Wong *et al.* argued, builds on the ability to scan information and adjust learning. These skills are particularly relevant in environments of ambiguity and uncertainty where "typical strategic situations lack structure, are open to varying interpretations, and potentially pertinent information is often far-flung, elusive, cryptic, or even contradictory."[26] Consequently, effective strategic leaders will be able to know which decision factors are most important in relation to the big picture; can identify root causes quickly and prioritize alternatives; integrate information from a variety of sources; and detect trends, associations and cause-effect relationships.

3. Cross–cultural Savvy includes the ability to understand cultures (of coalition partners as well as target populations) beyond one's own organizational, economic, religious, societal, political, and geographic boundaries. Especially as the frequency of coalition operational experiences is likely to increase in the future, the ability to understand and anticipate the values, customs, norms, and assumptions of other groups, organizations, and nations is becoming ever more important.

4. Interpersonal Maturity refers to the ability to display compassion and share power with subordinates, peers, and constituents. These traits are prerequisites for effective consensus and coalition building and for managing change proactively by embedding their vision within the organization and by shaping organizational culture to support that vision.

5. World-class Warrior means that strategic leaders move beyond tactical and operational competence and understand the entire spectrum of operations at the strategic level including theater strategy; campaign strategy; joint, interagency, and multinational operations; "and the use of all elements of national power in the execution of national security strategy."[27]

6. Professional Astuteness develops officers into leaders in their profession providing that the military will retain its special calling as the institution that serves the national defense and will not morph into "just another job, organization, bureaucracy or occupation."[28] Essential to the promotion of professional astuteness is institutional flexibility in allowing each individual to find intrinsic satisfaction in one's own self-concept as an officer as well as his or her individual acceptance of the Army profession's ethic and its place within American society.[29]

The need for strategic decisionmaking at every turn of an ever-more complex operational environment will require that these meta-competencies will be ingrained into the readily available skill set of military leaders and will become second-nature to their decisionmaking. How can this be accomplished? A brief excursion into the cognitive processes that determine decisionmaking under conditions of uncertainty might be instructive in this context.

DECISIONMAKING UNDER UNCERTAINTY

In an ideal world, we would all make our decisions based on an ordering of all alternatives and then base our choice in a rational manner on the alternative(s) that maximize expected utility (see Figure 1). Of course, in real life we do not have perfect information and cannot base our choices on decision strategies reflecting unbounded rationality. Instead, political scientist Herbert Simon convincingly demonstrated that people typically possess uncertain information about all their potential alternative choices and dispose only of limited computational capacity to determine their maximum utility function. To account for those limits of rationality, Simon suggested replacing the aim of maximizing an objective function with the more realistic concept of satisficing.

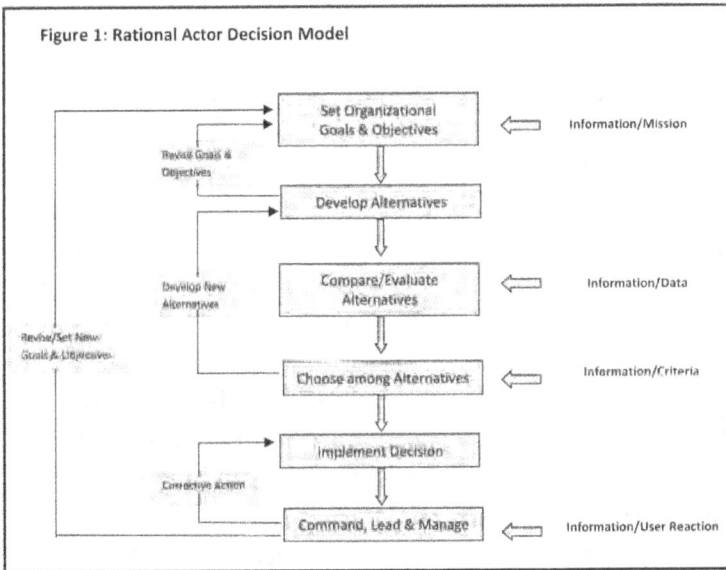

Figure 1: Rational Actor Decision Model

Set Organizational Goals & Objectives — Information/Mission

Revise Goals & Objectives

Develop Alternatives

Develop New Alternatives

Compare/Evaluate Alternatives — Information/Data

Revise/Set New Goals & Objectives

Choose among Alternatives — Information/Criteria

Implement Decision

Corrective Action

Command, Lead & Manage — Information/User Reaction

Figure 1. Rational Actor Decision Model[30]

Satisficing denotes "problem solving and decision making that sets an aspiration level, searches until an alternative is found that is satisfactory by the aspiration level criterion, and selects that alternative."[31] In other words, individuals create a threshold which allows them to demarcate their choices, accepting only alternatives above the threshold. Furthermore, ordering is no longer necessary, since individuals tend to choose the first alternative above the threshold, as that meets their requirements (which determine the threshold in the first place). But how are these requirements determined? What processes enable individuals to establish decision thresholds in the first place?

Psychologists have studied the way individuals make decisions in the presence of great uncertainty or incomplete information and have found that they often rely on mental shortcuts — called "heuristics" — to help them "reduce the complex tasks of assessing probabilities and predicting values to simpler judgmental operations."[32] Heuristics can be considered "rules of thumb," educated guesses, intuitive judgments, or simply common sense that are learned and honed by experience. More precisely, heuristics reflect strategies using readily accessible, though loosely applicable, information to control problem-solving in human beings and machines.[33] Although reliance on heuristics provides effective rational guidance in most circumstances, in certain cases it leads to systematic errors or cognitive biases that may skew decisionmaking.

Heuristics relevant to decisionmaking under conditions of uncertainty (and with relevance to strategic decisionmaking in complex contexts) include:[34]

- Anchoring and Adjustment. People start with an implicitly suggested reference point (the "anchor") and adjust their decisions based on that specific data point. However, anchoring may result in a focusing effect in that people place too much importance on one aspect of an event, thereby causing an error in accurately predicting the utility of a future outcome.[35]

- Representativeness. In many situations, an event A is judged more probable than an event B whenever A appears more representative than B. For instance, large samples are typically judged more representative than small ones, because their "salient features" or "essential properties" are thought to better reflect those of the population. Relying on the representativeness of an event as an indicator of its probability may lead to systematic errors in judgment, since it may either give undue influence to variables that affect the representativeness of an event, but not its probability, or it may reduce the importance of variables that are crucial to determining the event's probability, but are unrelated to the event's representativeness.

- Availability. Psychological experiments have revealed that whenever some aspect of the environment is made disproportionately salient or available to the perceiver, that aspect is given more weight in causal attribution. Thus, stereotypes, for example, can function as simplifying decision-guides to shape reality, or occurrences of extreme utility or disutility may appear more likely than they actually are, since people are typically preoccupied with highly desirable outcomes (e.g. winning the lottery), or with highly undesirable outcomes (e.g. an airplane crash). Kahneman *et al.* also found evidence of selective observation, i.e., people tend to perceive support for their initial beliefs, even if the evidence at hand disconfirms these beliefs. Thus, an individual may view his or her personal choice as less deviant—and more typical—than someone else's conflicting choice. In other words, the availability bias can skew the retrieval process itself, which, in turn, may yield an unrepresentative data base.

- Affect. Under conditions of uncertainty, feelings such as fear or pleasure may solicit an emotional response to the contextual stimulus. Affect enables us to make quick decisions and helps us avoid dangerous situations. However, our use of emotions to make decisions can also easily cloud judgment. For instance, in cases when an emotional reaction like fear is especially strong, it can completely overwhelm our reasoning process. Fear, Al Gore argues in his latest book, is the most powerful enemy of reason, citing the fact that almost three-quarters of all Americans were so easily led to believe that Saddam Hussein was personally responsible for the attacks of 9/11, and that many Americans still believe that most of the hijackers on 9/11 were Iraqis.[36]

Strategic decisionmaking in today's complex environment resembles conditions of uncertainty and rapid change. The role of the strategist "is to exercise influence over the volatility, manage the uncertainty, simplify the complexity, and resolve the ambiguity, all in terms favorable to the interests of the state and in compliance with policy guidance."[37] As a result, decisionmakers will rely more or less heavily on heuristic decision-rules in the absence of easily transferable prior experiences, applicable standard operating procedures, clearly defined rules of engagement, or rational decision calculi. Any curriculum intended to teach strategy and strategic decisionmaking must not only account for these cognitive shortcuts, it ought to take advantage of them by honing skills that enhance individuals' ability to quickly recall and employ desired heuristic decision patterns.

STRATEGIC DECISIONMAKING

Strategic decisionmaking in the future requires teaching meta-cognitive skills that provide leaders with a tool bag of decision options to use when confronting novel situations. This also requires the development of innovative and adaptable decision models beyond the rational actor model that has characterized traditional strategic decisionmaking (see Figure 1). The rational actor model is based on three main assumptions, all of which claim universal validity:[38]

1. Order — there are discoverable underlying cause-effect relationships in human interactions, the understanding of which in past behavior enables us to define "best practice" for the future.

2. Rational decisions — our choices are rational results of calculations of expected utility based on the desire to maximize pleasure and minimize pain.

3. Intentional capability — the mere acquisition of capability indicates automatically an intention to use that capability.

The rational actor model encourages/teaches individuals "to frame problems, formulate alternatives, collect data, and then evaluate options."[39] But the strategic context of complex environments demands creative and flexible decisionmaking not limited only to the rational application of predetermined rules and learned response patterns. The Department of Defense (DoD) 2002 Millennium Challenge wargame showed the limitations of doggedly applying the rational actor decision model to emerging complex situations (see Box 1).[40] Strategic decisionmaking in the future must be proactive and decentralized. Experience alone no longer adequately prepares leaders to be effective strategic decisionmakers, as situational awareness, cross-cultural considerations (in terms of organizational as well as international cultures), and trustworthiness are central skills to be applied to rapidly changing and increasingly complex decision contexts.

In 1995, Lieutenant General Paul Van Riper took a group of Marines to the New York Mercantile Exchange, "because the jostling, confusing pits reminded him of war rooms during combat. First, the Marines tried their hand at trading on simulators, and to no one's surprise, the professionals on the floor wiped them out. A month or so later, the traders went to the Corps's base in Quantico, Virginia, where they played war games against the Marines on a mock battlefield. The traders trounced them again—and this time everyone was surprised."[41] Analyzing the humbling results, the Marines concluded that "the traders were simply better gut thinkers. . . . They were far more willing to act decisively on the kind of imperfect and contradictory information that is all you ever get in war."[42] The traders, so Kurtz and Snowden opined, "were skilled at spotting patterns and intervening to structure those patterns in their favor."[43]

In the summer of 2002, the United States Joint Forces Command conducted the Millennium Challenge 2002 (MC02), at a price tag of $250 million the largest and most costly wargame exercise in American history. MC02 combined live field exercises and computer simulation in an attempt to test the military's transition toward new technologies that would enable network-centric warfare and provide more powerful weaponry and tactics. The game simulated war between the United States, denoted "Blue," and an unknown adversary in the Middle East, denoted "Red."

Red, commanded by retired Marine Corps Lt. General Paul K. Van Riper, used unconventional methods, including motorcycle messengers to transmit orders to front-line troops, evading Blue's sophisticated electronic surveillance network. Moreover, on the first day of play, Red launched a preemptive attack using air-, sea-, and ground-launched cruise missiles to sink 16 American ships. The equivalent of this success in a real conflict would have resulted in the death of more than 20,000 service personnel.

Unhappy with this unexpected result, the commanders decided to refloat the American ships and restart the game following predetermined plans of action. In response, Van Riper resigned from his role in the midst of the game, contending that the rest of the game was scripted for American victory. In an interview, Van Riper explained the peril of placing too much faith in technology at the expense of a deeper understanding of the nature of war:

War is about adapting. Any potential enemy as well as we, the United States, if we didn't adapt, learn, and evolve from our past experiences, we would be a species or a nation that would not survive. And any enemy that wants to survive against the United States can't fight like some of our recent enemies have, or they won't survive.

But just because the United States has overwhelming forces (or at least we Americans perceive that it does) and will for the foreseeable future, shouldn't make us believe that we're always going to dominate on that future battlefield. Many enemies are not frightened by that overwhelming force. They put their minds to the problem and think through: how can I adapt and avoid that overwhelming force and yet do damage against the United States? We've seen some of that in the latter stages of the war in Iraq, where an enemy that was defeated in a conventional battle is using some of the same techniques that the United States saw in Vietnam. . . .

There were accusations that Millennium Challenge was rigged. I can tell you it was not. It started out as a free-play exercise, in which both Red and Blue had the opportunity to win the game. However, about the third or fourth day, when the concepts that the command was testing failed to live up to their expectations, the command then began to script the exercise in order to prove these concepts. . . .

I think one of the fundamental lessons that should have been learned from Afghanistan is the ability to understand another culture. As Americans we're sort of arrogant in many, many ways about other cultures. We don't study them, we don't appreciate them. If we'd gone in there with units on the ground who didn't appreciate the culture, who couldn't immerse themselves in it and adapt to it, we'd have had a lot different outcome than we did. So if there's anything we need to look to in the future beyond continuing to develop the technology, it is to understand how we want to fight, and to become much more aware of the various regions and peoples of the world—how they think, how they understand the world, and how we relate to them.
Source: www.pbs.org/wgbh/nova/wartech/nature.html.

BOX 1. The Millenium Challenge.

ORDER AND UNORDER

Humans tend to employ heuristic shortcuts and other cognitive patterns to order the world and make sense of new situations. However, as the Millennium Challenge and the Marine-trader examples illustrate, complex decision contexts do not always lend themselves to patterned behavior, predetermined choices, or predictable outcomes. Ordered contexts allow us to rely on pre-established reductionist patterns focusing on efficiency in problem solving. In "unordered" contexts, "every intervention is also a diagnostic, and every diagnostic an intervention; any act changes the nature of the system."[44] Clausewitz refers to this phenomenon as "friction." Van Riper explains:

> You have the element of friction on the battlefield, for example. You can't account for friction. It just occurs. It's everything from a fuel tank that leaks and causes an airplane or a vehicle not to be able to perform its function, to an accidental discharge that a young soldier makes, to weather conditions. All of these have an interplay that causes the friction that leads to uncertainty.[45]

Kurtz and Snowden illustrate unorder by comparing it to the evolution of cities, in that "the two primary versions of urban arrangements, the planned and the 'organic,' often exist side-by-side. . . ."[46] In complex decision contexts, formal command structures and standard operating procedures tend to be complemented by informal trust networks. And in some circumstances, "'cultural factors,' 'inspired leadership,' 'gut feel,' and other complex factors are dominant."[47]

Recognizing the complexities of strategic decision contexts, Kurtz and Snowden developed a sense-making framework, labeled with the Welsh word *Cynefin*.[48] The Cynefin framework suggests four basic approaches to strategic decisionmaking, depending on the level of contextual uncertainty (see Figure 2):

1. Known (ordered): cause and effect relationships are generally linear, empirical and nondisputable; repeatability generates predictive models; focus is on efficiency and single-point forecasting, field manuals, operational procedures are legitimate and effective; structured techniques are mandatory.

2. Knowable (ordered): stable cause and effect relationships exist but may not be fully known; at issue is whether time and resources allow a move from knowable to known; decision model senses and analyzes incoming data and responds accordingly; structured techniques are desirable, but assumptions must be open to challenge; entrained patterns are most dangerous since simple error in assumptions may lead to false conclusion.

3. Complex (unordered): studies how patterns emerge through interaction or different agents; emergent patterns can be perceived but not predicted ("retrospective coherence"); decision model creates probes to make patterns of potential patterns more visible prior to taking action; understanding requires gaining multiple perspectives on the situation.

4. Chaos (unordered): no visible/perceivable cause-effect relationships; little to no response time; patterned responses may contribute to the chaos; decision model requires quick and decisive action to reduce the turbulence and then sense immediately the reaction to the intervention and respond accordingly.

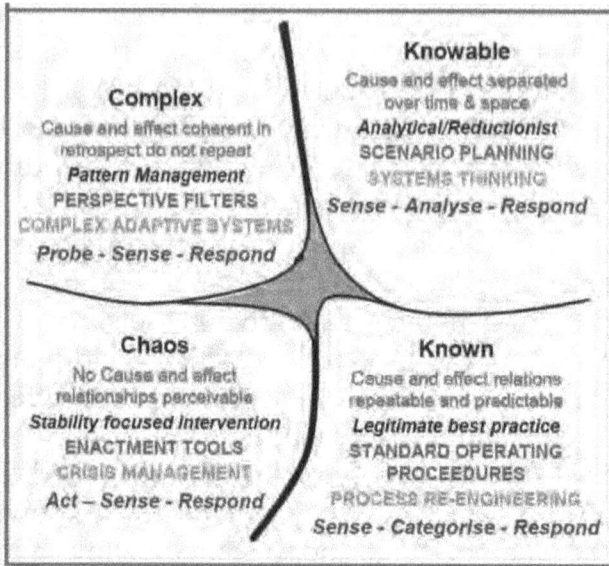

Knowable
Cause and effect separated
over time & space
Analytical/Reductionist
SCENARIO PLANNING
SYSTEMS THINKING
Sense - Analyse - Respond

Complex
Cause and effect coherent in
retrospect do not repeat
Pattern Management
PERSPECTIVE FILTERS
COMPLEX ADAPTIVE SYSTEMS
Probe - Sense - Respond

Chaos
No Cause and effect
relationships perceivable
Stability focused Intervention
ENACTMENT TOOLS
CRISIS MANAGEMENT
Act – Sense - Respond

Known
Cause and effect relations
repeatable and predictable
Legitimate best practice
STANDARD OPERATING
PROCEEDURES
PROCESS RE-ENGINEERING
Sense - Categorise - Respond

Figure 2. The Cynefin Framework.[49]

While the Cynefin framework seems to reflect the complex decision context of today's strategic environment, it is a very new conceptual approach that has not yet been tested thoroughly in military strategic decisionmaking contexts. Nevertheless, its great strength lies in the fact that it enables us to move beyond reductionist decision models designed for success in an ordered world. The primary challenge in applying the Cynefin framework to complex decision contexts will be to teach strategic decisionmakers—or teach them to discover—recognizable patterns (although their details may remain unpredictable), "stabilize or disrupt them depending on their desirability, and seed desirable patterns by creating attraction points."[50] And even in chaotic contexts, by recognizing that we don't know, we can begin to search for patterns and react to them. Using case studies can aid this process and sharpen the cognitive skills to make sense of complex or chaotic decision contexts.

CASE METHODOLOGY

When the Harvard Business School was started, it became apparent almost instantaneously that there were no textbooks suitable to graduate studies in business. Faculty members quickly set out to remedy this shortcoming by interviewing leading business people and writing detailed accounts of what they were doing. Of course, these first case studies could not yet reflect practices to be emulated, because there had not been any established criteria for determining success, effectiveness, or lessons learned. So the professors instructed their students to read the cases, to come to class prepared to discuss them, and to offer recommendations for appropriate courses of action. The case study methodology was born.[51]

A case tells a story. It recounts real or realistically simulated events, problems, or decision dilemmas so that students can wrestle with the complexities, ambiguities, and uncertainties confronted by the real or fictional decisionmakers. Case studies typically place students at the center of difficult decisions and illustrate to them how theory can be useful in addressing real world policy/decision dilemmas. More specifically, cases compel students to:

- distinguish pertinent from peripheral infor- mation,
- identify problems, dilemmas, decision para- meters and alternative courses of action,
- determine possible solutions,
- formulate strategies and policy recommen- dations, and

- recognize and confront obstacles to their implementation.[52]

Typically, cases are historical or retrospective, fictional or decision-forcing. Retrospective cases present a comprehensive account of a problem in history, specifying the actors involved and their positions and contending interests, the cycle of events, and the real outcome. Students are typically asked to analyze why certain decisions were taken and the observed result(s) obtained and to identify alternative options that may have led to a different outcome.

Decision-forcing cases stop short of revealing the outcome. Instead, they force students to get inside the heads of the decisionmakers — or the story's protagonist(s) and antagonist(s) —, wrestle with their decision choices and assess the utility of possible options for action. Decision-forcing cases usually include an epilogue that tells the rest of the story, i.e. what happened after the decision point with which the case leaves the reader. Students again, analyze why what happened, happened. Fictional cases usually revolve around events or decision dilemmas approximating real world policy problems.

Irrespective of type, cases typically highlight dilemmas at two levels: (1) pertaining specifically to the story of the case (case dilemma), and (2) pertaining to its larger policy implications (policy dilemma). Effective cases do not provide specific policy recommendations or definite answers for how to resolve the presented dilemmas. Quite the contrary, they present evidence in support of both (or more) sides of a policy argument and will often leave readers with some discomfort in terms of how dilemmas should be resolved. The absence of a one-sided argument, specific policy recommendations, or logical conclusions sets case studies apart from typical academic publications.

The purpose of using case studies in the classroom is to engage students in active learning and enable them to recognize the importance of the issues at hand as well as their greater policy implications. For maximum effectiveness, students should be asked to relate each case to other course materials or educational or professional experiences and to discern lessons that apply beyond the case context to other areas of their professional engagement, i.e., strategic thinking, decisionmaking, and leadership.

Teaching Cases.

Good case studies should be accompanied by detailed teaching notes providing instructors with a pedagogical road map for how to teach the case. While the specifics regarding the content of the case and its dilemmas and decision points will vary, a few guiding principles apply to using case studies in any classroom setting. Case teaching is a discursive endeavor. Students should be guided but not told. Discussion replaces lecture and the primary role of the instructor is the facilitation of that discussion.

Typically, students will begin by identifying the primary policy dilemma(s) presented in the case. Next, they should discuss its/their major implications (e.g., for U.S. national security policy) and, based on other course readings, recommend a specific course of action for guiding future decisions in similar situations. Most generally, instructors may want to start the classroom discussion by asking a series of general questions (these could pertain to any case) including:

- What is this case about?

- What is the chronology of events?
- What is/are the problem(s) and dilemma(s) specific to the case?
- What are the larger policy implications exemplified by this case?
- Who are the main actors (i.e., persons, agencies, states, organizations, companies, etc.) that make up the story? Who is the protagonist, and who is the antagonist?
- What are their respective positions/interests? How do they pursue those interests?
- For professional/military students:
  - How are this case, the dilemmas and implications relevant to your experiences and organization?
  - How do they affect how you do your job?
  - How do they challenge your management and leadership capabilities?

These general questions are then followed by more targeted questions pertaining to the specific case and policy dilemmas illustrated in the case. Case teaching differs from seminar discussions of academic arguments. Students need to understand that cases and their underlying dilemmas usually do not have one right solution. Therefore, students are not expected and should not attempt to find the best recommendation(s). Instead, their task is to clearly identify the case and policy dilemmas and ponder their implications. From their analysis, they should derive logical recommendations for how problems could be avoided, challenges met and dilemmas solved in similar circumstances in the future.

CONCLUSION: TEACHING STRATEGY USING CASES

 Depending on pedagogical objectives, we need to distinguish between (a) the use of case studies in the classroom to illustrate a series of topical issues or dilemmas and have students wrestle with their solution and (b) the use of the case methodology for honing skills, developing effective heuristic shortcuts and cognitive response mechanisms, or shaping behavior patterns. Referring back to the quadrants of the Cynefin sense-making framework, the use of case studies in traditional classroom settings has proven a valuable tool for teaching the known and analyzing the knowable. Yet contemporary strategic decision environments more often than not require skilled responses to complex and at times chaotic situations. Applying learned rigid, rational actor-type decision rules under conditions of great uncertainty will likely render suboptimal results, as in the example of the Millennium Challenge. Instead, strategic decisionmaking in complex operations will call on individuals to rely on a combination of experience, skill, speed, creativity, adaptability, and intuition. No two situations are exactly alike. To make sense of novel situations, we rely on mental shortcuts. Exposure to new situations — real or simulated — will hone heuristic skills.

By using case studies in a specific course, we supply not only the simulated decision context but also — oftentimes unintended — the heuristic frame within which dilemma solutions are derived. The course content — topics, readings, prior discussions, etc. — increases the salience of certain issues (availability bias), thereby providing an indirect frame steering the discussion and solutions in a particular direction.[53]

Unfortunately, the complexities of the contemporary strategic environment cannot be addressed by a few classes. Strategic decisionmaking does not happen in 90-minute sessions on Tuesdays and Thursdays. This insight has significant implications for teaching strategy.

Instead of teaching strategy in one or a series of dedicated classes, PME curricula ought to be interfused with strategic decision choices. In addition to learning how the military develops strategy and derives strategic decisions (using the rational actor model), the curriculum ought to be structured in such a way that students can also hone heuristic and intuitive skills. Specifically, this would mean focusing on learning to recognize/perceive emergent patterns, respond to them and quickly assess, and if necessary correct, the course of action. Using case studies — and simulation exercises — frequently will help hone these skills. Adopting case studies on nonmilitary topics illustrating dilemmas with little or no connection to national security — e.g., management, trade, development, public policy, or business related cases — will force students out of their professional comfort zone and challenge them to look for emerging patterns, move away from predetermined thresholds, and reflect on the implications of alternative decision choices.

Teaching strategy effectively means stimulating students continuously to "get inside the heads" of case protagonists with widely differing cultural backgrounds, professional experiences, individual and organizational interests — e.g., by identifying with the contextual demands placed on different government agencies, foreign leaders, nongovernmental organization (NGO) representatives, rebel force commanders, or civic leaders. Selecting and using case studies with this objective in mind will enhance students' understanding of the cognitive frames of other actors relevant to a specific decision context and directly help to hone the strategic leadership competencies outlined above. Important in this context will be repeated exposure to unorder, challenging students to make a choice between "allowing the entrained patterns of past experience to facilitate fast and effective pattern application and gaining a new perspective because the old patterns may no longer apply."[54]

Strategy in the traditional sense is about control: control over means and ends and over the resources to achieve them. In the military, strategy has been aimed at controlling VUCA. But our inherent desire for control fails in decision contexts characterized by complexity or chaos, as illustrated by the Millennium Challenge wargaming exercise. Managers and by extension strategists can learn effective response patterns under conditions of unorder from the way we manage children: "they use boundaries and interventions to encourage desirable behavior but do not attempt to control it through goal-based direction."[55]

Teaching strategy under conditions of complexity and chaos means fostering and encouraging continuous learning and innovation by internalizing a process of sense-making through pattern recognition and instinctive, adaptive responses. Consequently, teaching strategy must be more than simply training decisionmakers in the science of calculating "objectives, concepts, and resources within acceptable bounds of risk to create more favorable outcomes than might otherwise exist by chance or at the hands of others."[56] It must also be about the art of understanding complexity and recognizing the value and interaction of order and unorder. Teaching strategy is about effectively using heuristic decision devices to make sense of new situations and recognize the possibilities for shaping them in a desired direction. This means, honing the skills necessary for making sense of chaos, including the recall of heuristics that worked well in the past. Exactly here is where teaching strategy may employ case studies most effectively.

ENDNOTES - CHAPTER 9

1. Carl von Clausewitz, *On War*, Radford, VA: Wilder Publications, 2008.

2. This line is widely attributed to Benjamin Disraeli, although the exact circumstances in which he said it is unknown. Available from *www.quoteopia.com/quotations.php?query=life*.

3. Headquarters, Department of the Army, *Field Manual (FM) 3-07, Stability Operations*, Washington, DC: U.S. Government Printing Office, 2008.

4. Headquarters, Department of the Army, *Field Manual (FM) 1-1, The Army*, Washington, DC: U.S. Government Printing Office, 2005, p. 1-1.

5. FM 3-07, preface.

6. See also National Security Presidential Directive NSPD-44, available from *www.fas.org/irp/offdocs/nspd/nspd-44.html*; and DOD Directive 3000.05, available from *www.dtic.mil/whs/directives/corres/pdf/300005p.pdf*.

7. See for instance Gabriel Marcella and Stephen O. Fought, "Teaching Strategy in the 21st Century," *Joint Force Quarterly*, Issue 52, No. 1, 2009, pp. 56-60.

8. Clausewitz, p. 147.

9. *Ibid.*

10. Stephen A. Shambach, *Strategic Leadership Primer*, 2nd Ed., Carlisle, PA: U.S. Army War College, 2004.

11. Basil H. Liddell Hart, *Strategy*, 2nd Rev. Ed., New York: Praeger, 1968.

12. *Ibid.*

13. *Ibid.*, pp. 348-349; a synopsis is available from *home.att. net/~nickols/articles.htm.*

14. See Fred Nickols, "Strategy: Definitions and Meaning," available from *home.att.net/~nickols/strategy_definition.htm.*

15. Harry R. Yarger, *Strategic Theory for the 21st Century: The Little Book on Big Strategy*, Carlisle, PA: Strategic Studies Institute, U.S. Army War College, February 2006, p. 5, available from *www. strategicstudiesinstitute.army.mil/pubs/display.cfm?pubID=641.*

16. In this context, see also Kenneth Andrews, *The Concept of Corporate Strategy*, 2nd Ed., Dow Jones Irwin, 1980, p. 4.

17. Nickols, "Strategy: Definitions and Meaning," p. 5.

18. See, for instance, Samuel Huntington, *The Soldier and the State*, Cambridge, MA: Harvard University Press, 1957; Morris Janowitz, *The Professional Soldier: A Social and Political Portrait*, Glencoe, IL: The Free Press, 1960; Volker Franke, *Preparing for Peace*, Westport, CT: Praeger Publishers, 1999; and Peter Feaver and Richard Kohn, eds., *Soldiers and Civilians: The Civil-Military Gasp and American National Security*, Cambridge, MA: MIT Press 2001.

19. *Ibid.*, p. 7.

20. Yarger, *Strategic Theory for the 21st Century*, p. 65.

21. Robert H. Dorff, "A Primer in Strategy Development," in Joseph R. Cerami and James F. Holcomb, Jr., eds., *U.S. Army War College Guide to Strategy*, Carlisle, PA: Strategic Studies

Institute, U.S. Army War College, 2001, p. 14, available from *www. strategicstudiesinstitute.army.mil/pubs/display.cfm?pubID=362*.

22. *Ibid.*, p. 18. See also *www.au.af.mil/au/awc/awcgate/stratdm. htm*.

23. See Leonard Wong, Stephen Gerras, William Kidd, Robert Pricone, and Richard Swengros, *Strategic Leadership Competencies*, Carlisle, PA: Strategic Studies Institute, U.S. Army War College, September 2003, available from *www.au.af.mil/au/awc/awcgate/ssi/ ldr_comps.pdf*.

24. *Ibid.*, p. 5. These lists include, for instance, the Army's Strategic Leadership Primer, an updated 2004 version of which is, available from *www.au.af.mil/au/awc/awcgate/army-usawc/sprimer. pdf*; and *Field Manual (FM) 22-100, Army Leadership*, available from *usacac.army.mil/CAC/CAL/FM6_22.pdf*.

25. William Steele and Robert Walters, "21st Century Leadership Competencies: Three Yards in a Cloud of Dust or the Forward Pass?" *Army Magazine*, August 2001, p. 31, quoted in Wong *et al.*, *Strategic Leadership Competencies*, p. 6.

26. Wong *et al.*, *Strategic Leadership Competencies*, p. 6.

27. *Ibid.*, p. 9.

28. Already in the late 1970s, military sociologist Charles Moskos warned that with the creation of the All-Volunteer Force in 1973, the U.S. military had begun transitioning from an institution to an occupation in the eyes of many of its members. See Charles Moskos, "From Institution to Occupation: Trends in Military Organization." *Armed Forces & Society*, Vol. 4, No. 1, 1977, pp. 41-50.

29. See Wong *et al.*; and Don Snider, "The [Missing] Ethical Development of the Strategic Leaders of the Army Profession for the 21st Century," conference paper, XIII Annual Strategy Conference, Carlisle, PA, April 10, 2002.

30. The figure was adopted from Fred W. Nickols, "Strategic Decision Making: Commitment to Strategic Action," available

from *home.att.net/~essays/strategic_decision_making.pdf*. A detailed discussion of the rational actor model can be found in Graham T. Allison and Philip Zelikow, *Essence of Decision: Explaining the Cuban Missile Crisis*, 2nd Ed., New York: Longman, 1999.

31. Herbert Simon, *Models of Bounded Rationality*, Cambridge, MA: MIT Press, 1982, p. 168.

32. Daniel Kahneman, Paul Slovic, and Amos Tversky, *Judgment under Uncertainty: Heuristics and Biases*, Cambridge, MA, New York, and Melbourne and Sidney, Australia: Cambridge University Press, 1982, p. 3.

33. See, for instance, Judea Pearl, *Heuristics: Intelligent Search Strategies for Computer Problem Solving*, New York: Addison-Wesley, 1983, p. vii.

34. See Kahneman *et al.*, *Judgment under Uncertainty*.

35. For example, Kahneman *et al.* demonstrated that when asked to guess the percentage of African nations that are members of the United Nations, people who were first asked "Was it more or less than 45 percent?" guessed lower values than those who had been asked if it was more or less than 65 percent. This pattern held in other experiments for a wide variety of different subjects of estimation. See *Ibid.*

36. Al Gore, *The Assault on Reason*, New York: Penguin Press, 2007. Excerpts are available from *www.abcnews.go.com/GMA/Story?id=3198208&page=1*.

37. Yarger, *Strategic Theory for the 21st Century*, p. 18.

38. For further detail, see, for instance, Cynthia F. Kurtz and David J. Snowden, "The New Dynamics of Strategy: Sense-Making in a Complex and Complicated World," *IBM Systems Journal*, Vol. 42, No. 3, 2003.

39. Thomas A. Stewart, "How to think with your Gut," Business 2.0, November 2002, p. 2, available from *www.cognitiveedge.com/ceresources/articles/49_Thinking_with_your_Gut_(_T_Stewart_article_in_Bus_2).pdf*.

40. For further detail, see *web.archive.org/web/20070928005405/* *www.jfcom.mil/about/experiments/mc02.htm*; Van Riper interview, available from *www.pbs.org/wgbh/nova/wartech/nature.html*.

41. *Ibid.*, p. 2.

42. *Ibid.*

43. Kurtz and Snowden, "The New Dynamics of Strategy," p. 466.

44. *Ibid.*, p. 466. In contrast to directed or designed order, Kurtz and Snowden refer to "emergent orders" as unorders, which does not depict a lack of order but rather a substantively different kind of order.

45. PBS Interview, available from *www.pbs.org/wgbh/nova/* *wartech/nature.html*.

46. Most historic towns, and virtually all those of metropolitan size, are puzzles of premeditated and spontaneous segments, variously interlocked or juxtaposed." Quoted in Kurtz and Snowden, "The New Dynamics of Strategy," p. 466.

47. *Ibid.*

48. *Ibid.*, p. 467. Kurtz and Snowden explain:

> The name Cynefin is a Welsh word whose literal translation into English as habitat or place fails to do it justice. It is more properly understood as the place of our multiple affiliations, the sense that we all, individually and collectively, have many roots, cultural, religious, geographic, tribal, and so forth. We can never be fully aware of the nature of those affiliations, but they profoundly influence what we are. The name seeks to remind us that all human interactions are strongly influenced and frequently determined by the patterns of our multiple experiences, both through the direct influence of personal experience and through collective experience expressed as stories.

49. *Ibid.*, p. 468. See also Nickols, "Strategic Decision Making."

50. Kurtz and Snowden, "The New Dynamics of Strategy," p. 481.

51. The Harvard Business School began using case studies in its instruction in 1925. Detailed information on the program is available from *harvardbusinessonline.hbsp.harvard.edu/hbsp/case_studies.jsp.*

52. Detailed information on the case methodology and its use in the classroom can be found in Vicky Golich, Mark Boyer, Patrice Franko, and Steve Lamy, "The ABCs of Case Teaching," Pew Case Studies in International Affairs, Washington, DC: Institute for the Study of Diplomacy, Edmund A. Walsh School of Foreign Affairs, Georgetown University, 2000, available from *www.usc.edu/programs/cet/private/pdfs/abcs.pdf.*

53. For instance, in my Introduction to Political Science course, I assign William Golding's *Lord of the Flies.* Most students have read this book in English class in high school. In my class, however, the reading of the book is preceded by readings by Aristotle, Hobbes, and Marx. The entire frame with which students read or reread a well-known story changes to reflections about governance and human nature.

54. Kurtz and Snowden, "The New Dynamics of Strategy," p. 467.

55. *Ibid.*, p. 481.

56. Yarger, *Strategic Theory for the 21st Century*, p. 1.

CHAPTER 10

TEACHING STRATEGY IN 3D

Ross Harrison

THE CASE FOR TEACHING STRATEGY[1]

History abounds with brilliant strategists who never undertook formal academic training in the subject of strategy. Napoleon Bonaparte and Winston Churchill on the military and diplomatic fronts, and Michael Dell, Warren Buffett, and Bill Gates from the world of business, all distinguished themselves by developing and executing brilliant strategies without any formal training as strategists. Should we then conclude from this that strategy is an intuitive skill that gifted individuals are born with, rather than something that can be taught? Or is it, in fact, possible to train individuals to think and act strategically through rigorous academic discipline and training?

Great historical strategists notwithstanding, I believe that strategy is a mental discipline that can, in fact, be developed in astute individuals willing to submit to a rigorous academic process. Though undoubtedly individuals may be born with innate mental capacities that predispose them to be master strategists, strategic aptitude consists of a cluster of skills, such as integrative thinking, risk assessment, and situational analysis, all of which can be developed through a combination of academic and practical training. This in no way should be construed as meaning that innate qualities are not helpful in the development of the strategist. It is only to suggest that there are "strategies for teaching strategy"[2] that can be quite effective with individuals who might not be so naturally endowed.

This having been said, there is no silver bullet method for teaching strategy. One reason for this is that strategy is not a skill like basic accounting that can be simply conveyed, but rather is a higher level discipline that must be rigorously developed in students over time. There really is no linear method for teaching students how to develop and execute strategy, much like there is no singular path for teaching students how to write or analyze. Instead, students must develop the competency and build the mental circuitry of strategic thinking over time and under strong academic stewardship.

Moreover, because the real world is by nature untidy and imprecise, there are great challenges in developing a pedagogic methodology for strategy which closely approximates this reality. To be effective, approaches to teaching strategy must accommodate decisionmaking within complex, changing, and oftentimes inscrutable real world environments. Adding to the complexity is the fact that thinking and operating strategically require more than just intellectual prowess. Since strategic environments are by nature dynamic and fluid, consummate strategists must also be emotionally equipped to accommodate changes in reality, particularly when they contradict intensely held preconceptions.

The traditional way of teaching strategy is to focus students on the lessons of the great strategists rather than on the development of strategic thinking. As Gabriel Marcella suggests, traditionally we teach about strategy, rather than how to develop strategy.[3] That is, we study the great strategists on the assumption that somehow the travails of the masters might hold some universal strategic truths for us.[4] Though a comprehension of the experiences of consummate strategists is undoubtedly important to an understanding of strategy, more is required than merely assimilating the experiences of others. In addition to incorporating the historical lessons of the masters, we need to clearly define strategy and then derive from this a methodology for developing strategic thinking skills.

One question that needs to be addressed is whether there is a one-size-fits-all method for teaching strategy, or whether the training of strategists must take place in application specific contexts. An argument might be made that since there is significant variance in how strategy is conducted in military, foreign policy, and business milieus, any pedagogic approach that treats strategy as fungible across these different arenas is fundamentally flawed. While it is true that the differences in how strategy is devised among these disparate fields are profound and students should ultimately study strategy within the context of their chosen field, there are fundamental skills involved with strategic thinking that can be conveyed in a cross-disciplinary way. If we keep in mind that our initial task is not to train students how to conduct specific strategies, but rather first to develop strategic thinking skills, we can more easily discover some of the universal elements of strategy.

But these obstacles and caveats notwithstanding, it is possible to develop an approach for teaching strategy. What is necessary to effectively train students in strategy is to model and simulate strategic situations as closely as possible, and then compel them to weigh alternative strategic options within these contexts. Though strategy is as much an art as it is a science, with no real simple axioms or how-to's, there are effective pedagogic methods for developing strategic capacity in students.[5]

HOW TO TEACH STRATEGY IN 3D[6]

The Challenge of Defining Strategy.

One obstacle that must be overcome before students can start down the path of developing strategic thinking skills is the demystification of the concept of strategy. The fact that strategy has become part of the popular parlance certainly does not make our task any easier. Strategy is commonly used to connote something as simple as "a good idea," or something as prosaic as "a plan of action."[7] Even some of the academic and professional literature on strategy contributes to the mystification of the concept. Much of the literature distills strategy down to a simplified, direct relationship between means and ends, mysteriously omitting any real analytic treatment of what this relationship entails.[8] Though it is true that strategy does involve the nexus between means and ends, a useful definition of strategy has to go beyond this overly linear, black-box depiction.

Before we can embark on the journey of teaching students how to think strategically, we have to define strategy in terms that can be operationalized. Strategy is a set of mutually reinforcing actions, calculated to have a compounding effect on an organization's environment, on behalf of its goals. That is, strategy entails creating advantage by concentrating resources and actions towards shaping an organization's environment. Though this definition does not contradict the notion that strategy entails a nexus between means and ends, it does move us away from treating it as a linear connection, and shifts our focus to the important role situational environments play in the means-ends relationship.

Thinking in 3D.

Developing a methodology for teaching strategy involves expanding on the definition mentioned above. As our definition implies, strategy cannot be conceptualized in a vacuum, but rather only within the context of an organization's situational environment. It follows from this that one of the keys to teaching strategy involves students becoming adroit at analyzing and modeling these environments. The primary thesis of this chapter is that the most effective way to do this is to compel students to think about situational environments in 3-dimensional terms. This method is derived from the insight that strategic success and failure oftentimes hinge on outcomes at three interconnected dimensions of an organization's environment. Therefore, strategy is like playing three games simultaneously, each within a different dimension of this environmental reality.[9] What follows from this is that strategic advantage oftentimes comes from mutually reinforcing wins in all three dimensions of the environment. And conversely, strategic failure can come from compounding, mutually reinforcing losses within these same three dimensions. Because of this, teaching students to conceptualize environments in 3D is an indispensable part of the strategy training process.

So what are these three critical dimensions of strategic environmental reality that exist with almost all organizations? They are the dimensions of systems, actors, and targets. Each of these will be explained below, and then later we will use these as our backdrop for thinking about and teaching strategy in 3D. To ground us in what otherwise might become an overly abstract exercise, we will draw from multiple business and foreign policy cases as we progress.

When we think about strategy, competitive situations are what immediately come to mind. Whether we are thinking within a business, foreign policy, or military context, we normally conceive of strategy as something which creates a decisive advantage over one or more particular adversaries.[10] But this normal instinct notwithstanding, students need to model the systems dimension first. The reason for this is that competitive dynamics for most organizations take place within a broad systems context, something the strategist needs to take into account before proceeding further. Moreover, strategic opportunity or threat can in fact be posed by the nature and structure of these systems. For business environments, the systems dimension consists of industries, while for foreign policy it consists of international systems of sovereign states and multilateral organizations.

Though the systems dimension is one of the most important for students to model, it is also the most abstract and oftentimes the most difficult to grasp. And, in fact, it is frequently overlooked because it is not as concrete as focusing on the particular characteristics of individual competitors. But this systemic dimension is essential, as it is the primary backdrop for a country's grand strategy and many a global company's corporate strategy. In both of these cases, strategy involves targeting not just specific adversaries, but also the broader structural context within which competition with adversaries takes place. Part of business strategy involves influencing or adapting to the structure of industries, as even dominant companies are susceptible to the gravitational forces of these powerful systems. Similarly, foreign policy strategists need to account for the influence of regional and international systems on state behavior. Failure to incorporate these systems into their strategic calculus is done at the business leader's or policymaker's peril.

The second dimension of the environment that must be addressed by students is the dimension of actors. Here, the emphasis is on the attributes of individual competitors, allies, and one's own organization, rather than on the systemic context. Since strategy entails thriving in an environment of competing wills and intellects, addressing competitive dynamics and the particularities of individual actors is incredibly important. In particular, students and practitioners of strategy need to be adept at analyzing the capabilities and the motivations of their most formidable adversaries, their most fervent allies, and their own organizations. In the actors dimension for both business and foreign policy, the focus is on studying, modeling and assessing the nature of the adversary or ally, and then developing strategic responses that appropriately take these into account.

The third dimension of the environment that needs to be addressed is the dimension of targets. This captures the ground level at which much of competitive strategy takes place. But the emphasis here is not on individual actors; rather it is on the target groups whose support is essential for strategic success. For business, the dimension of targets is represented by markets, while for foreign and military policy it is one's own domestic support base or the mass publics of an ally or adversary. The presumption here is that with foreign policy strategy, competitive advantage necessitates targeting action and resources towards shaping groups of mass publics, while successful business strategy requires the targeting of market groups or segments.

The danger in treating the system, actor, and target dimensions in isolation from one another is that students may fail to see how they are interconnected. That is, though it is important to conceive of these dimensions as discrete, they are, in fact, different facets of the same environmental reality. The systems dimension forms the broad context for strategy, while the actors dimension consists of competitors and allies, and the targets dimension is comprised of groups that need to be targeted for strategic success. Together, these interrelated dimensions comprise the strategic environment for businesses, governments, and militaries. So the key to any pedagogic approach to strategy is training students to think and operate simultaneously within all three of these dimensions. Since strategic success is oftentimes created by compounding wins at all three of these dimensions, it becomes clear that teaching strategy in a 3D context is incredibly important.

Acting strategically in 3D.

Once students become adept at thinking in 3D, they are ready to practice developing strategy within this multidimensional context. As part of this process, it is critical that students understand the relationship between strategic action and the situational environment. One useful device for thinking about the causal connection between strategic action and outcomes in the environment is the distinction between unitary and bundled strategies. Unitary strategies consist of a single action designed to have an impact on one or more dimensions of the environment, while bundled strategies consist of multiple actions, each targeted at a different dimension of the situational environment. As we will see later, some of the most powerful strategies are these complex bundled strategies which involve simultaneous or sequential launchings of actions aimed at each of the three dimensions of the environment. Businesses, for example, may aim action at specific competitors (actors), but as part of the same strategy may also aim at industry structure (systems), and attempt to influence trends in their markets (targets). Similarly, foreign policy strategy may involve actions aimed at specific states (actors), but may also target the structure of international or regional systems, as well as attempt to influence mass publics (targets). So, these bundled strategies do not consist of single actions; but rather they embody a central concept that acts as a directional beacon, pointing out the general pathway for a set of mutually reinforcing actions aimed at different dimensions of the environment.

Now that we have reinforced the notion of strategy operating in multiple dimensions, we need to take a more granular look at strategy within each of the three dimensions of the situational environment. Later we will see how students need to put this all together and learn to play simultaneous games, each at a different dimension of the situational environment.

Strategy Within the Systems Dimension.

Adapting To or Shaping the System? It is important for students to understand that one of the first decisions that must be made is whether the direction of a given strategy will be to adapt to or shape the system.[11] In other words, is the purpose of a given strategy to operate within or change the status quo? Niche market strategies for companies and political-military containment strategies for countries generally fall within the domain of adaptive strategies, while innovation strategies for companies and roll-back military strategies for countries would generally be categorized as shaping strategies. As useful as this distinction between shaping and adapting strategies is, at times the line between them becomes blurred. For example, strategy may catalyze systemic change that is already evolving. In cases like this there is a bit of both adapting to and shaping the system. That is, though the strategic intent may be to shape the system, what in fact takes place is the strategist exploits change that is already afoot, pushing the system to a "tipping point."[12] One could argue that the Reagan administration's strategy vis-à-vis the Soviet Union was such a tipping point strategy. The logic was that the Soviet Union was over-extended, collapsing slowly under its own weight. What the Reagan administration did, according to this line of thinking, was to catalyze that trend, thereby accelerating the Soviet Empire's demise.

Strategies Targeting System Functions. Regardless of whether one is dealing with adaptive or shaping strategies, it is important to map the various functions that make the targeted system work. The reason for this is that strategy in this dimension may be aimed at either adapting to or shaping the functional make-up of the system. Some systems like the international financial system are complex and may be highly differentiated functionally, while other systems such as small, newly emerging industries may be less differentiated and not nearly as complex. The key is to be able to identify the various functions and then map how they connect together to make the system work. In business cases, students of strategy need to map the various functions contained within an industry's value chain, such as distribution channels, manufacturing, supply chains and logistics.[13] Though international political systems may not be as clearly defined as business industries, they too may be functionally differentiated. Students mapping the international system during the Cold War could identify different actor types, such as superpowers, client states, uncommitted states, and supra-national actors, each serving very different functions within the bipolar, balance-of-power system.[14]

Another system characteristic that needs to be modeled involves the roles played by key actors, particularly those playing the role of disrupter of the status quo, at one end of the spectrum, and those acting as regulators, at the other. For example, when looking at the Middle East system, one could argue that Saudi Arabia, despite its occasional rhetoric to the contrary, plays the role of regulator and protector of the regional status quo. On the other hand, al-Qaeda, and some might argue Iran, plays the role of disrupter of regional and international status quos. In the business world, one could argue that Microsoft plays the role of regulator in the software industry, while Yahoo and Google play the role of disrupter.[15]

So, how does the student of strategy use this modeling of functions and roles to tease out strategic possibilities? The idea is that the modeling exercise will highlight opportunities for the strategist-in-training to adapt to, or alternatively, shape the functional make-up of the system. That is, students will look for opportunities to create strategic advantage by adding, eliminating, or changing some of these functions. As an example, Michael Dell clearly understood how the personal computer industry was functionally organized. He observed that the profitability of PC manufacturers was getting squeezed at two ends of the industry's value chain, both by the supply chain and distribution channel functions. On the supply chain side, Intel and Microsoft were increasingly putting pressure on industry margins; and on the distribution channel side, the mass retailers were also flexing their buying power muscles and putting downward pressure on profits. Given the resultant lackluster performance of most PC producers and waning industry economics, Dell saw the futility of trying to adapt his company to the existing industry structure. So instead he targeted the functional structure of the industry itself, disintermediating the traditional distribution channel, and sold directly to the end-user. His company's strategy involved disrupting the system by changing its functional make-up, thereby capturing profits that otherwise would have accrued to the mass retailers.[16]

On the foreign policy side, the Obama administration's nascent strategy for Iran can be thought of as seeking to disrupt the functional makeup of the Middle East system by luring Syria away from its alliance with Iran. The idea is to neutralize the functional role Syria serves as intermediary between Iran and its Arab proxies of Hezbollah and Hamas. The concept is that the administration's bargaining position on the Iranian nuclear issue would be enhanced by eliminating the function that Syria serves as Iran's conduit to radical causes in the region.

This notion of Syria serving an intermediary function for Iran's connection to the broader region can also be useful in a military strategy context. In the eventuality of a military confrontation with the United States, Iran could launch a counteroffensive using Syria as a conduit for mobilizing Hezbollah and Hamas against U.S. and Israeli interests in the region. But a U.S. strategy of luring Syria away from an alliance with Iran could change the military landscape by attenuating Iran's ties with Hezbollah and Hamas, circumscribing the country's capacity to project power into the Mediterranean. Military strategists in both Washington and Tehran will likely factor this into their respective strategic analyses and military planning processes.

Strategies Targeting System Linkages. Most political, foreign policy, and business systems are not completely discrete, but rather are linked to other systems of activity. One industry may be linked to another related industry, and a regional political system, for example, may be linked to other regional systems and the national political system. The student of strategy needs to be aware of these linkages, as oftentimes they provide the opportunity for strategic advantage. Focusing on linkages also makes reification of systems less likely, as it points to the arbitrary nature of system boundaries. Students in particular need to be focused on strategies which aim to break down the boundaries between linked systems — the result being the birth of new systems. For example, Apple Computer's strategy with its path-breaking iPhone was to act as a system disrupter, breaking down the barriers and creating convergence between the mobile phone, entertainment, and computer industries. Prior to the commercialization of the iPhone and other similar PDA devices, these industries were viewed as discrete and disconnected. It was Apple who envisaged the linkages and aimed to destroy the barriers between these three separate industry systems. The result was the meteoric rise of Apple and the creation of an entirely new industry.

The notion of system linkages is an incredibly important concept for students to understand, as it forces them to think outside current structures and see possibilities for creating strategic change. It also forces students to look out for vulnerability points in systems that might be suggestive of new and creative strategic possibilities.[17] Moreover, it compels students to think in integrated terms, something that is important for all aspects of strategic thought and action.

Strategies For Coping With System Change.

One of the most daunting aspects of formulating strategy is coping with system change. As much as it might be the fantasy of students and practitioners of strategy to freeze their situational environments in time, systems are in a constant and dynamic state of flux. But a limitation of some of the methods for modeling these environments is a failure to capture change, many merely providing a snapshot of the system at a fixed point in time. So herein lies the conundrum. If strategic success is premised upon a given situational environment, how can winning strategies be developed within environments that are in a constant state of change? Compounding the problem, how does the strategist account for the impact that their strategy will have on system change in the future?[18]

Though coping with system change can be difficult, there are some ways to prepare students in this area. One method is to sensitize students to emerging trends, such as whether the system is tending towards fragmentation or consolidation, or how the distribution of power is changing. Here the student extrapolates from current trends where the system is seemingly headed. In some circumstances, this trend analysis can be an effective way to deal with system change.

But with most system change there is a residual amount of uncertainty that cannot be adequately captured with trend analysis or extrapolation. Fortunately there are some analytic methods for coping with varying degrees of uncertainty about system change. One of the most compelling methodologies for doing this has been developed by Hugh Courtney, who posits four levels of residual uncertainty, ranging from a single clear view of the future that is extrapolated from current trends, to a limited number of possible future outcomes, to a bounded range of future outcomes, to an unlimited (and ambiguous) range of possible futures.[19] He proposes different types of strategies, both of the adapting and shaping varieties, for coping with these varying degrees of uncertainty about the future. Hedging and probing represent a couple of adaptation methods for coping with uncertainty, while innovation and controlled system disruption represent some useful shaping strategies. Another methodological variant that students can learn from is the scenario planning approach pioneered by the oil giant, Royal Dutch Shell, which essentially tests the robustness of different strategies against different possible alternative views of the future.[20]

Military strategists also have techniques available for coping with system change and uncertainty. Wargaming, which tests the robustness of specific military strategies against multiple political and military scenarios, is one method that is commonly used. Wargaming forces players to make decisions within environments fraught with uncertainty, and also compels them to face the reality that their strategies may change the environment in ways that further increase uncertainty.[21] The use of intelligence also plays a role in reducing uncertainty and mitigating risk in the military context. Intelligence analysis can be used to glean information about an adversary's likely moves or about conditions in the broader environment. For example, during the Cold War, the U.S. Central Intelligence Agency used information gleaned from economic espionage conducted against the Soviet Union, to conclude that this adversary was grossly overextended and that its military expenditures were unsustainable.[22] This calculus and the reduction of uncertainty that came from the use of intelligence, emboldened the Reagan administration in its posture vis-à-vis this historic adversary.

Strategy within the Actors Dimension.

While creating strategy within the systems dimension of the situational environment provides the broad structural context for strategy, students also need to learn to develop strategy within the dimension of actors, which include both competitors and allies. While there is a paucity of literature in the area of systems strategy, the field of competitive strategy is much richer, replete with work on all aspects of competitive foreign policy, military, and business strategy.[23] Much of the research on strategy, in fact, focuses on this competitive dimension, where the emphasis is on competing wills and intellects and on the attributes of specific actors.[24]

Before developing competitive strategy, students must learn to model the actor dimension by identifying and then conducting analysis of key adversaries and allies, as well as one's own organization. The focus of the analysis should be on two key attributes of actors: capability and motivation. Capability points to the range of options an actor has available, and the resources it has at its disposal. Motivation helps us understand how an actor is likely to make decisions within this range of options.

To start an analysis of capability, students need to be able to make the distinction between absolute and relative capability.[25] While absolute capability addresses the specific power attributes of an actor, relative capability is more useful for strategic analysis because it measures how an actor's overall power compares with that of other actors in the environment. The reason this is important is that it forces the student to consider the possibility that an adversary's absolute capability may remain constant, while its power relative to other actors in the environment may increase or decrease. For example, it has been well-documented that Iran's bargaining leverage vis-à-vis the United States was significantly enhanced by the invasion of Iraq in 2003. What changed for Iran was not its absolute capability, but rather its relative capability. That is, with the collapse of its historical nemesis in Iraq, Iran's capability relative to the United States, and to its regional neighbors, spiked significantly.

In addition to relative capability, there are other factors that need to be considered in an analysis of the capability of an adversary, an ally, or one's own organization.[26] One of these factors is the type of strategic assets the organization has in its arsenal. Students learn to distinguish between hard and soft strategic assets. While hard assets are finite and tangible, like military ordnance, technology, and financial capital, soft assets are less tangible, like human capital, management, and leadership. Any analysis for an adversary, ally, or one's own organization needs to take into account both hard and soft strategic assets as part of the capability calculus.

Another critical capability factor that students of strategy need to be sensitive to is the complementarity of strategic assets. Though taking an inventory of individual hard and soft strategic assets is necessary, it is insufficient. Since strategic capability is more than the sum of individual assets, it is also important to consider how these assets work together and reinforce one another. The most successful organizations develop a repertoire of complementary strategic assets. Amazon, the giant online retailer, uses the combination of technology, leadership, marketing savvy, and its alliances with publishers to generate its strategic capability. Their tremendous success and capabilities can only be explained by the complementary nature of these different assets, and how they have been made to work in tandem.

Another important dynamic that students need to consider as part of any capability analysis is the distinction between deployable and bargaining assets. That is, some assets are most effective when they are actually deployed, while others are most effective when kept on the sidelines to yield bargaining leverage. During the Cold War, the United States used both deployable and bargaining assets. It deployed conventional weaponry and other material support to our allies, but used its nuclear weaponry capability as an instrument of bargaining leverage and to create deterrence. All organizations, including businesses, governments and militaries have a mixture of both deployable and bargaining assets. A capability analysis for any of these organizations needs to factor in these different types of strategic assets.

The second important factor in the actor dimension that students need to analyze is motivation. As students learn, strategic success oftentimes hinges on accurately assessing whether a given adversary is motivated primarily by threat or opportunity. This will indicate whether a competitor is likely to operate from an offensive or defensive posture, a data point that can be used to predict how the adversary will respond to your strategy.[27] It can also be useful in pointing to likely decisions that an actor will make within its range of available options. Since strategy at many levels entails competition between competing wills, an understanding of the motivations of the opponent is an important part of any competitor analysis.[28]

After conducting capability and motivational analyses for adversaries, allies, and one's own organization as well, the next step is to develop competitive strategy within this context. As part of this training, students are taught the distinction between offensive and defensive types of competitive strategies. Though presenting an exhaustive list of offensive and defensive competitive moves is outside the purview of this chapter, there are a few that are particularly worthy of mention. On the offensive strategy side, base attack and root attack strategies are incredibly important. Base attack strategies aim at the operational core of an adversary. An example of a base-attack strategy is the Bush administration's military attack on al-Qaeda's bases of operation in Afghanistan. Another example would be Allied bombing raids on Germany's steel making capacity during World War II. Root attack strategies go deeper in that they aim to neutralize an adversary by targeting their sources of power. Root attack strategies focus less on the adversary's base of operations, but more on interdicting the power sources that supply that base. An example would be the current U.S. strategy of using drones to attack targets in western Pakistan in an attempt to cut al-Qaeda off from its critical power assets. Targeting al-Qaeda's funding sources in Saudi Arabia and elsewhere represents another example of such a strategy. In addition to base and root attack strategies, pressure point strategies involving attacks on high value unprotected flanks of an adversary and the strategy of surprise represent other types of offensive competitive strategies.[29]

In terms of sensitizing students to the art of offensive strategy, some of the work from the world of business is incredibly useful. Michael Porter's delineation of competitive strategy into three types; differentiation, cost, and focus, is a helpful framework for students. That is, companies can compete on the basis of distinguishing themselves from the competition, having a lower cost structure than the competition, or focusing laser-like on a given product category or market segment.[30] But Porter's framework also has some relevance for foreign policy and military strategy. Countries and armies, too, can create strategic advantage by creatively differentiating themselves in terms of capabilities from their adversaries, or by acting more efficiently and effectively. For example, Porter's strategy of focus is similar to Liddell Hart's concentration of military power, discussed in his path breaking treatise on strategy.[31]

In addition to offensive competitive strategies, students should also be schooled in defensive strategy. Deterrence and containment strategies on the foreign policy and military fronts and barrier-to-entry strategies in business are examples of defensive strategic moves. Though Israel's military and political strategies clearly embody offensive components, one could argue that they use a defensive strategy of deterrence to dissuade the country's adversaries from attacking. Similarly, companies strive to create sustainable strategic advantage by building barriers-to-entry, oftentimes using operational scale as a kind of deterrent mechanism to discourage would-be competitors from entering their markets.

Strategy within the Targets Dimension.

The part of the situational environment missed in most analyses of strategy is the targets dimension. But strategists who omit consideration of this dimension are woefully remiss, as this comprises a critical element of successful strategy. As mentioned before, for foreign policy these targets are groups consisting of one's own mass publics or those of an adversary, while for business they are market groupings of existing and potential customers. Within this dimension, governments aim to mobilize support for their foreign policies, and businesses aim to generate demand for their products and services.[32]

We even see these types of strategies being used in the military arena, where building political support among target groups for military endeavors can be essential. General David Petraeus's strategy of co-opting the Sunni insurgents in Iraq is a prime example. His approach, which proved to be at least tactically successful, improved the U.S. military's competitive advantage vis-à-vis al-Qaeda by building support among Iraq's previously hostile Sunni groups.[33] Also, revolutionary movements, and even some terrorist organizations like al-Qaeda, target groups of masses using nationalist or religious symbols as part of a strategy of mobilizing opposition to existing political and social systems.[34]

So given the importance of this dimension, how can the student model it? The main factor that has to be accounted for is the structure and organization of target groups. Is the market or the body politic highly differentiated or largely monolithic? If it is differentiated, how is it structured and organized? For governments, this means identifying the most important political groups, while for businesses it means delineating how their markets are segmented. Also relevant are the trends in terms of growth or decline among these groups. That is, which groups are gaining strength in numbers and power, and which are on the decline? Success of U.S. presidential campaigns often hinges on how astutely a given candidate's team identifies groups that coalesce or diverge around specific issues. And business success is determined by how well companies identify and then target segments of customers with similar tastes, behaviors, or attributes. Since strategic success often depends on how well this is done, it is important for students to grasp this type of analysis.

So how do students use the analysis of groups to learn to create winning strategies, and what types of strategies do students need to be exposed to so as to be effective within this dimension? On the business side, companies use market segmentation as a strategy to increase demand for their products or services, by identifying groups of customers with similar needs, behaviors, or attributes. They then build demand by targeting products, services, and brands at these segments. The retailer Target, for example, created demand through a strategy of identifying a new market segment: affordable fashion. While high end retailers like Nieman Marcus and Nordstrom's target the fashion segment and discounters like Wal-Mart targeted the affordable segment, Target created a new hybrid segment consisting of both affordable and fashion, giving the chain incredible strategic advantage.[35]

Another type of target strategy is using innovation to foster growth in target groups. Apple Computer, for example, expanded the size of the market for portable music with the launch of its innovative iPod. Being the innovator, Apple created a new market for MP3 players, and after becoming the market maker focused its strategic energy on stimulating demand and growing the size of this market. And one could argue that U.S. President Lyndon Johnson's "Hearts and Minds" campaign aimed at the Vietnamese people during the Vietnam War was a similar type of strategy.[36] Though the upshot was strategic failure, his idea was to win the support of the local populace to improve the U.S. position vis-à-vis the Vietcong.

Teaching Strategy in 3D: The Role of Case-Studies, Simulations, and Internships.

Up to this point, we have discussed how students deal sequentially with the three dimensions of strategy. This is important, as the student needs to work one-dimensionally before they can operate in 3D. But if the key to teaching strategy is ultimately having students think and act multidimensionally, then it is important that they eventually graduate from the sequential approach and become facile at conducting simultaneous strategies in 3D. There are case studies available from both the foreign policy and business worlds which are useful in this regard, requiring students to think and make decisions multidimensionally.[37] These cases can be used for in-class exercises, or in simulations that compel students to role-play in interactive, dynamic strategic situations.

One of the most useful cases I have found for building the mental circuitry required for thinking and acting in 3D is the Dell Computer Company.[38] The case is also incredibly helping in reinforcing the concept of bundled strategies. As mentioned before, Dell's strategy of selling direct to the customer consisted of a cluster of individual actions, each creating mutually reinforcing wins within the three dimensions of the company's situational environment. The result was that Dell was able to gain a compounding advantage by generating multiple wins in the system, the actor, and the target dimensions.

So this case is important not only because it helps students understand Dell's wins within each dimension of the company's environment, but because it also helps them see the compounding effect of mutually reinforcing wins at multiple levels. Let's briefly recount the case, outlining each of these mutually reinforcing wins.

Win 1-System Strategy. Students will see that Dell's strategy for the system dimension was to disintermediate the retail channel, and thereby usurp profits from the mass-retailer. This system level strategy enabled Dell to turn an unfavorable industry environment into a virtual goldmine of profits for the company. This win reversed the poor industry economics, and allowed Dell to operate profitably. It also gave the company the financial resources to pursue the other two legs of its strategy.

Win 2-Actor Strategy. Another leg of Dell's strategy was conducted within the actor dimension of its situational environment. Even while Dell was operating

298

his business out of a college dorm room, he specifically targeted IBM customers. His competitive strategy was to differentiate by offering customized machines at lower cost to the customer. He correctly assessed that IBM would underestimate the challenge coming from an upstart entrepreneur, and be slow to respond. Dell's competitive strategy of cost and differentiation gave his company the ability to build an initial core customer base of medium and large businesses on the backs of IBM and other competitors.

Win 3-Target Strategy. But Dell's success would not have been as decisive as it was had it been content to merely differentiate its machines from IBM and cut the mass-retailer out of the industry's value chain. The most compelling move, and the company's secret sauce, was its target strategy. This consisted of Dell restructuring the market, thereby increasing overall market demand for personal computers. To do this, they deployed a strategy of market segmentation powered by an incredible data-base management capability. Prior to Dell's market ascent, the market had been organized into large, amorphous segments, consisting of business and governmental institutional customers and consumers. What Dell did was deploy a scalpel-like segmentation strategy, using its data management skills to slice and dice the market into increasingly granular segments, and then offer customized product to each of these micro-segments. The result was enhanced customer loyalty, an increase in market demand, and an incredible growth trajectory for Dell.

This case shows students how a bundled strategy helped Dell create compounding strategic advantage out of mutually reinforcing wins. They see that Dell's system strategy of disintermediation of the retailer gave the company access to a large portion of the computer industry's profit pools, and enabled it to get closer to the customer (WIN 1). Students also witness how its competitive actor strategy of focusing on customized solutions allowed it to differentiate itself from IBM in meaningful ways, and later the other competition (WIN 2). And they see how its winning target strategy of micro-segmentation engendered growth in the market and unleashed an avalanche of demand for Dell computers (WIN 3). This case demonstrates to students how these three mutually reinforcing wins gave Dell the compounding strategic advantage it enjoyed for years.

There are also foreign policy cases which provide some of the same pedagogic opportunity for teaching strategy in 3D. The Obama administration's nascent strategy for Iran, for example, could be a useful scenario for a capstone simulation in a course on strategy. In contrast to the emerging Obama strategy, the Bush administration's strategy for Iran had been largely conducted within the competitive actor dimension. Its strategy for getting Iran to renounce its nuclear path consisted of threats of regime change, the specter of military action, and interdiction of Iran's banking relationships. However, since Iran's strategic environment had improved over the years since the U.S. invasion of Iraq, this competitive actor strategy alone proved to be ineffective. What the Obama administration seems to be heading towards is a systems dimension strategy of diminishing Iran's capacity to project power into the Mediterranean by luring Syria away from its alliance with Iran. President Obama also seems to be trying to bring Russia into the anti-nuclear fold in an attempt to isolate Iran and improve the competitive U.S. bargaining leverage.

And as we saw recently, President Obama is using his immense personal popularity to target a populist message at the Iranian people in a way that might put pressure on and circumscribe the options of the ruling clerics in Tehran.

One might argue that these types of case studies and simulations bear little resemblance to the way strategy is actually conducted in the real world. Though this argument might have some validity, it misses the point. The intention is not to suggest that this 3D pedagogic approach exactly replicates how strategy is practiced in all domains. Rather it is to propose this method as a heuristic device for building strategic thinking and decisionmaking skills. Moreover, these types of simulations and exercises that model strategic situational environments are similar to what is already being used by several U.S. Government agencies that are working on foreign policy and military strategic initiatives.[39] Additionally, the major petroleum companies, Royal Dutch Shell in particular, conduct some of the more dynamic and inventive situational modeling, testing their long-term strategic initiatives against dynamic environmental scenarios.[40] So though these exercises may not always precisely mimic reality, they can be incredibly useful in training students to think and act multidimensionally, the first step in the creation of strategists.

That having been said, efforts to train students in strategy should not end in the classroom. The use of case studies and simulations should be supplemented with internships and other experiences that immerse students in real life strategic situations. Internships with major consulting firms represent some of the best opportunities to reinforce what is learned about strategy in the classroom. One reason internships with large consultancies have been so effective is that students get exposure to a broad, cross-section of clients from both the private and public sectors, and get to work on a full range of strategic problems and opportunities. The advantage of working with such a broad range of clients and problems is that the astute intern will discern strategic patterns that can be applied to later situations and with possible future clients.

Another reason that internships with major consulting firms can be a powerful component of the strategy training process is that these organizations use a very effective hypothesis-based method for solving strategic problems for their clients.[41] The steps involved with this methodology: problem definition, diagnosis, hypothesis, and solution, all require rigorous analysis of the client's strategic environment, something that reinforces what students have learned in the classroom. Moreover, consultants learn to think multidimensionally, as success depends on their ability to test hypotheses against a fully developed view of the client's strategic environment. Student interns working in this environment, therefore, will burnish their skills of developing strategy within a 3D context. Also, the use of decision trees by consultancies as part of the problem solving methodology helps interns hone their decisionmaking skills. The consultant typically works with business and government clients towards making strategic decisions by helping articulate the assumptions about the environment that underpin each strategic option. Students fortunate enough to be witness to and part of this important process learn how strategic dilemmas are modeled, and become adept at navigating the analytic process that facilitates strategic decision. In sum, these internships can reinforce the 3D strategy skills students glean from the classroom experience, but also give them an opportunity to pick up other strategic skills that can only be acquired in a real world work setting.

CONCLUSION

Though what has been presented in the previous pages constitutes the core of a course on strategy with some recommendations for supplemental simulations and internships, students also need exposure to a broader academic curriculum that addresses other topics central to the development of strategic thinking. My earlier admonishment about the overuse of history in strategic training notwithstanding, a solid grounding in historical analysis should become part of any curriculum on strategy. Though strategy may be abstract and conceptual, it also needs to be rooted in the human experience. History is essential to understanding the experiences of failures and successes of past strategies. History is also critical to a study of leadership, a central component of all successful strategies. Political science is also an important dimension, particularly to the degree that it furthers an understanding of systems and power. Another area that needs to be incorporated into a curriculum on strategy is an integrated approach to risk analysis. The development of both quantitative and qualitative methods for assessing and managing risk is essential to weigh alternative strategic options. Courses which address decisionmaking and its centrality to strategy, such as economics and business management, are also helpful curricular additions.

If individuals undergo the kind of 3D training suggested here, will they necessarily become consummate strategists? There is obviously no certainty here, no more than a curriculum on writing is guaranteed to produce great writers. But the methodology does expose students to the essentials of strategy, such as 3D situational environmental analysis and contextualized decisionmaking. Though there are no certainties, treating strategy as an acquirable discipline rather than as a mysterious attribute, will lead us closer to developing strategic thinking skills and to our goal of creating competent strategists.

ENDNOTES - CHAPTER 10

1. I want to thank my friend, Hernan Narea, for his editing suggestions.

2. Gabriel Marcella, "The Strategy for Teaching Strategy in the 21st Century" *Of Interest*, Strategic Studies Institute, U.S. Army War College, November 8, 2007.

3. *Ibid.*

4. See Peter Paret's comments about the question of how useful Causewitz's work has been in influencing the practice of strategy in Peter Paret, ed., *Makers of Modern Strategy: From Machiavelli to the Nuclear Age*, Princeton, NJ: Princeton University Press, 1986, pp. 210-213.

5. Harry R. Yarger, *Strategic Theory for the 21st Century: A Little Book on Big Strategy*, Carlisle, PA: Strategic Studies Institute, U.S. Army War College, 2006, pp IX. Also see Harry R. Yarger, *Strategy and the National Security Professional: Strategic Thinking and Strategy Formulation in the 21st Century*, Westport, CT: Praeger Security International, 2008.

6. I want to thank Anthony Arend, Georgetown University Professor and Director of Georgetown's MSFS Program, for suggesting the use of the 3D metaphor.

7. Yarger, *Strategic Theory for the 21st Century*, p. v.

8. *Ibid*, p.1.

9. This is not to be confused with the more quantitatively rigorous notion of game theory.

10. I want to thank Chester Crocker, Georgetown Professor and former Assistant Secretary of State, for his comments about strategy taking place within the context of two or more competing minds and intellects. Also see André Beaufre, *Introduction a la Strategie [Introduction to Strategy]*, Paris, France: Economica, 1985, for a similar type of analysis.

11. Hugh Courtney, *20/20 Foresight: Crafting Strategy in an Uncertain World*, Boston, MA: Harvard Business School Press, 2001, Chapter 3.

12. Malcolm Gladwell, *The Tipping Point: How Little Things Can Make a Big Difference*, Boston: Little Brown and Company, 2000.

13. Michael Porter, *Competitive Advantage: Creating and Sustaining Superior Performance*, New York, NY: The Free Press, 1980. Porter first developed the concept of value chain in this book.

14. Morton A. Kaplan, *System and Process in International Politics*, New York: ECPR Classics, 1957. See Kaplan's discussion of essential national actors and supra-national actors in balance-of-power systems.

15. Kenneth N. Waltz, *Theory of International Politics*, New York: McGraw Hill Companies, 1979, provides a discussion of system dynamics and the roles of disrupters and regulators.

16. Jan Rivkin and Michael Porter, *Matching Dell*, Cambridge, MA: Harvard Business School Publishing, 1999.

17. See W. Chan Kim and Renne Mauborgne, *Blue Ocean Strategy*, Cambridge, MA: Harvard Business School Publishing, 2005, for an excellent analysis of how system barriers can be broken down to reveal new strategic possibilities. Also see James E. Moore, *The Death of Competition: Leadership and Strategy in the Age of Business Ecosystems*, New York, NY: Harper Collins, 1996,

for a discussion of strategic possibilities outside the traditional competitive realm.

18. See Clausewitz's concept of friction, which implies uncertainty in Paret, ed., *Makers of Modern Strategy*, p. 202

19. See Courtney, *20/20 Foresight: Crafting Strategy in an Uncertain World*, for systematized approaches for dealing with uncertainty about the future. As part of this, he posits the four levels of residual uncertainty. Also see Robert Rubin and Jacob Weisberg, *In an Uncertain World: Tough Choices from Wall Street to Washington*, New York: Random House Publishing: 2003.

20. Shell International BV, *Shell Energy Scenarios to 2050*, 4th Ed., The Czech Republic: The Hague, 2008.

21. See Mark Herman, Mark Frost, and Robert Kurz, *Wargaming for Leaders: Strategic Decisionmaking from the Battlefield to the Boardroom*, New York: McGraw Hill Companies, 2009.

22. See Patrick Cronin, ed., *The Impenetrable Fog of War: Reflections on Modern Warfare and Strategic Surprise*, Westport, CT: Praeger Security International, 2008, pp. 10-11, for a discussion of the use of intelligence to cope with the possibility of strategic surprise. Also see "At Cold War's End: U.S. Intelligence on the Soviet Union and Eastern Europe, 1989-1991,"Washington, DC: Central Intelligence Agency, CSI Publications.

23. See Carl Von Clausewitz, *On War*, Colonel J. J. Graham, trans., Wilder Publications, 2008, as an example of this type of work on competitive strategy.

24. I want to thank Georgetown University Professor Chester Crocker for his insights about competing intellects and adversaries being the crux of strategy. I also want to thank my colleague and friend, Thomaz Guedes da Costa, professor at the National Defense University, Washington DC, for his insights on the importance of the maneuver in competitive strategy.

25. SWOT analysis, pioneered in the 1960s and 1970s by Albert Humphrey for Stanford University, focused on a model of Strengths, Weaknesses, Opportunities, and Threats (SWOT), which implicitly dealt with the notion of relative capability.

26. I want to thank Rob Stein and Georgetown Professors Casimir Yost and Chester Crocker for their invaluable comments on the importance of self-assessment and inwardly directed strategies to the study of strategy.

27. See Porter, *Competitive Strategy*, chapter 3.

28. See discussion around Clausewitz's notion of the psychological/motivational dimensions of strategy in Paret, ed., *Makers of Modern Strategy*, p. 204.

29. See Chester A. Crocker's "Reflections on Strategic Surprise", *The Impenetrable Fog of War: Reflections on Modern Warfare and Strategic Surprise*, Patrick M. Cronin, ed., Westport, CT: Praeger Security International, 2008.

30. Porter, *Competitive Strategy*, chapter 2.

31. B. H. Liddell Hart, *Strategy*, Fourth Printing, New York: Praeger Publishers, 1975, chapter XX.

32. The activities of the U.S. State Department's Bureau of Public Affairs addresses this dimension on the foreign policy front.

33. For a comprehensive account of this strategy, see Thomas E. Ricks, *The Gamble: General David Petraeus and the American Military Adventure in Iraq, 2006-2008*, New York: Penguin Group, 2009.

34. See John Shy and Thomas Collier in Peter Paret, *Makers of Modern Strategy: from Machiavelli to the Nuclear Age*, Princeton, NJ: Princeton University Press, 1986, p. 815. Also in same volume, see Paret's discussion about Clausewitz's notion that military means are used on behalf of political objectives, pp. 206-207.

35. See Laura Rowley, *On Target: How the World's Hottest Retailer Hit a Bull's-eye*, New York: Wiley, John and Sons, Inc, 2003.

36. Lyndon Johnson's quote on May 4, 1965, reads: "So we must be ready to fight in Viet-Nam, but the ultimate victory will

depend upon the hearts and the minds of the people who actually live out there. By helping to bring them hope and electricity you are also striking a very important blow for the cause of freedom throughout the world."

37. For an anthology of foreign policy cases, see Ralph G. Carter, ed., *Contemporary Cases in U.S. Foreign Policy: From Terrorism to Trade,* Washington, DC: Congressional Quarterly, Inc, 2008.

38. The main case I use is Jan Rivkin and Michael Porter, *Matching Dell,* Cambridge, MA: Harvard Business School Publishing, 1999. Other Harvard cases on Dell that are useful are V. Kasturi Rangan and Marie Bell, *Dell – New Horizons,* Case 502-022, Boston, MA: Harvard Business School, 2002; Michael Dell and Joan Magretta, "Power of Virtual Integration: An Interview with Dell Computer's Michael Dell," *Harvard Business Review,* 1998; Hong Iris Wang, Ali Farhoomand, and Pauline Ng, "Dell: Selling Directly, Globally,"Boston, MA: Harvard Business School, 2007; and Ali Farhoomand and Mary Ho, "Dell: Overcoming Roadblocks to Growth," *Harvard Business Review,* 2006. Also for a case that highlights a strategy that was used against Dell, see Harvard Case: Yigang Pan and Kevita Sethi, "Lenovo: Countering the Dell Challenge," *Harvard Business Review,* 2005.

39. See *Desert Crossing Seminar: After Action Report June 28-30, 1999,* declassified by Brigadier General George J. Trautman, Jr., USMC Deputy 15, USCENTCOM.

40. Shell International BV Publication, "Global Scenarios: 1995-2020," available from *www.shell.com/home/content/aboutshell/our_strategy/shell_global_scenarios/previous_scenarios/previous_scenarios_30102006.html.*

41. See Ethan M. Rasiel and Paul N. Friga, *The McKinsey Mind,* New York: McGraw Hill, 2002, for a description of the McKinsey method of problem solving.

CHAPTER 11

BEYOND ENDS-BASED RATIONALITY: A QUAD-CONCEPTUAL VIEW OF STRATEGIC REASONING FOR PROFESSIONAL MILITARY EDUCATION

Christopher R. Paparone

The ends-based, rationalist model of strategy has dominated institutions of professional military education to the point it has become an ideology that limits the education of the professional military strategist. This chapter proposes a quad-conceptual framework that permits educators and practitioners of military strategy to contemplate simultaneous forms of reasoning beyond the ends-based rational model. This broadened perspective advocates a patterned view of strategic reasoning.

INTRODUCTION

The world is full of intractable situations and fraught with ambiguity. Some say that it has become increasingly so, but this has actually been the case all along and educators and practitioners of strategy just have the luxury of viewing the past through the lens of causal certainty, a lens that does not work when looking toward the future.[1] Their retrospective sense of certainty epitomizes the fallacy of the proverbial Monday-morning quarterback. Only through the study of history do they know how things ended up. Knowing how the story ended, institutions can attribute causal relationships that reinforce beliefs that such ends can be rationally achieved through purposeful strategies toward the future. Indeed, this knowledge of the past reinforces an ideological bent toward ends-based rationality; hence, provides the historic context for the objectification of an imagined future.[2] The inculcated belief is that the art and science of the military professional is to first understand undesirable situations as problems and then address them with envisioned clarity. Only now are the most reflective professional military education (PME) institutions discovering this rationalistic values system is losing relevance as practitioners continue to face highly intractable and ambiguous situations.

The military profession has relied too much on the expectations envisioned by the limited philosophy of ends-based, rationalistic models of strategy. As the profession struggles with making sense of complex, ambiguous world events, the end game view has produced false expectations. The hope of ends-based rationalism — to create effective strategies, plans, and decisions to reach a desired future end state — has been confounding. Yet, our PME institutions continue to teach this Weberian *Zweckrationalität* (sociologist Max Weber's term, meaning "ends-rationality") version of strategic thinking, assuming that practitioners can decide ahead of time how to employ resources in ways to achieve the ends we have in mind. The current PME system, that includes war colleges and staff colleges, is so infused with ends-based rationalism as to have unfortunately created a relatively closed ideology rather than a more open philosophy of reasoning. This chapter asserts that the institutionalization of ends-based rationalistic ideology has crippled the ability to educate practitioners who are neither exposed nor required to consider other philosophical forms of strategic reasoning. The idea here is not to disparage ends-based rationality by itself, but to subordinate this unitary form of reasoning to a more holistic view.

The purpose of this chapter is to challenge the prevailing PME unitary reasoning of ends-based rationality by exposing the practitioner to alternative views. In that regard, this chapter presents a meta-framework that subordinates the idea of ends-based rationality and claims that view must compete with other forms of reasoning contingent upon the degrees of tractability and ambiguity in the situation. This idea of quad-conceptual reasoning (thinking in fours) provides a diagnostic tool for practicing strategists to better make sense of situations through the lenses of other worldviews rather than just that of the ends-based rational model.[3] The challenge is to create a framework that enables practitioners to contemplate four forms of reasonableness — here strategy is defined as the development of theories of action through multifaceted contextualizations and recontextualizations of situations. In that regard, precursors to theories of action should be viable theories of reasoning (i.e., how we contextualize situations); hence, this is the primary argument of this chapter.

The proposed model of reasoning recognizes that there are varying degrees of tractability (tameness or manageability) present in the situations that practitioners face. The framework also considers levels of ambiguity (from purely objective accounts of reality to the most subjective ones). When these continua are crossed (tractable-unambiguous, and intractable-ambiguous), the resultant four sources of strategic reasoning: *programmatic* (appropriate for most tractable and least ambiguous situations); *planning* (for less tractable and lesser ambiguous situations); *participative* (for the more tractable, yet more ambiguous situations); and, *reflective* (for the least tractable and most ambiguous situations). The sections that follow will highlight these theories of reasoning, provide examples for each theory, and then discuss implications they each have for PME development (see Figure 1).

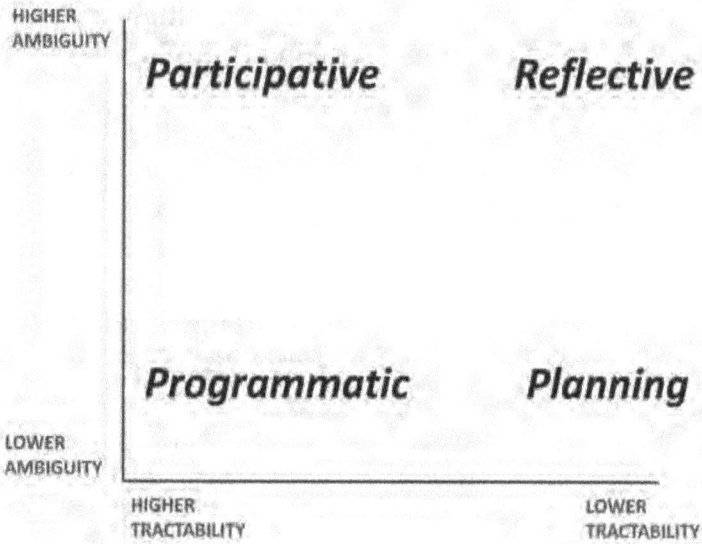

Figure 1. The Author's Model of a
Quad-Conceptual View of Strategic Reasoning.

With this proposed framework, defining strategy is revealed as an appreciative challenge. Each form of reasoning contemplates strategy with its own sense of reality.

STRATEGIC REASONING THROUGH PROGRAMMATICS

Theory of Reasoning.

For situations diagnosed as "tractable-objective," strategy is the best served with a programmatic logic. Here, practitioners view strategy as the structured process of employing technology to solve recurrent problems.[4] Another name for this form of ends-based reasoning is using "technical rationality" (i.e., strategies are pre-engineered solutions or technologies). Here, strategic problems are recognized with respect to what is known about solutions to recurrent problems; that is, there is a tight coupling between the process of problem identification and the solutions that already exist. This is the most extreme form of ends-based rationality because it assumes solutions are technically available (i.e., situations that draw attention become problems only when strategists see them as amenable to well-defined solutions).[5] Reasoning is a recognition and matching process.

As such, programmatic competence depends on technical rationality, defined as "the application of theories and techniques derived from systemic, preferably scientific research to the solution of the instrumental problems."[6] The strength of this form of reasoning is that education can be oriented on the known-knowns. Strategists can call upon technical expertise found in the hard science disciplines such as physics, systems engineering, operations research, computer science, and so on. Reasoning becomes a pairing process where situations are broken down into tractable problems which can be addressed with these proven techniques; hence, the problems are actually defined by the solutions. Programmatic strategic-reasoning requires an objective-view of reality, like considering the physical positioning of forces.

Indeed, a prime example of a programmatic strategy is that the United States has historically employed a *force-projection strategy*. The problem is objectively framed by the solution — when policymakers demand, the United States must be able to send and support forces over oceans (the solution is a force deployed through a transportation system). In order to move X amount of forces, we need Y number of trains, trucks, ships, airplanes calculated with respect to space and time. The underlying logic is technique based in mathematical models of space and time, hypothesis testing, breaking down and isolating variables (simplification), and so on. Factors of time, distance, tonnage, supply chains, and similar aspects for which the engineering sciences have an objective, cause-and-effect character about which the practitioner can be precise. In this sense, the field of military logistics is conceptually inseparable from the idea of military strategy (e.g., strategy is programmed with the creation, movement, and sustainment of forces along strategic lines of communication) (see Figure 2).

This diagram was created by the author to depict the phases of a force projection strategy, composed of a series of operations and activities that are pre-engineered for success. Here, the force projection process is a strategically reasoned technology.

For strategic weapons, like intercontinental ballistic missiles, the more the technological solution can stand off (in this case, there is no need for overseas land bases), the more the weapon system can be defined as strategic. In the 1950s, push-button warfare was the characterization of reasoning through programmatics. The dominant metaphor for this type of reasoning is indeed the machine.

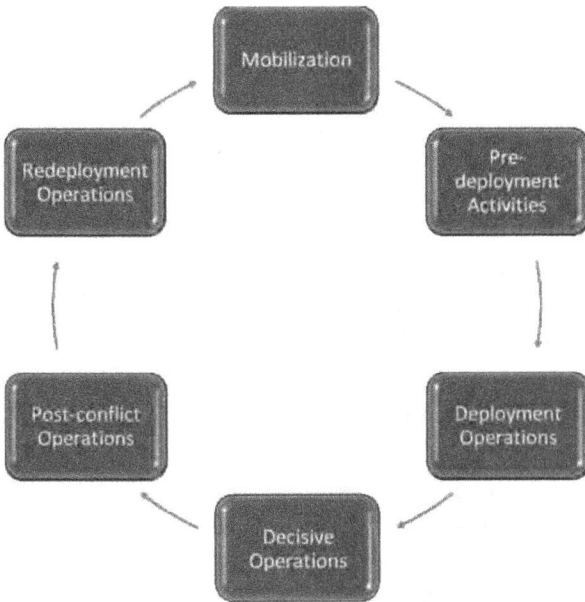

Figure 2. Strategic Reasoning Through Programmatics.

IMPLICATIONS FOR PME

There are important PME considerations in developing programmatic thinkers. Educating toward programmatic reasoning is supported by *assimilative knowledge*, or the science of attaching technical solutions to recurrent problems.[7] Assimilative knowledge can take the form of organizations (i.e., packages of technology) charged with building and performing machine-like recurring tasks, routines, standing operating procedures, doctrine, records, rules, tactics and procedures, textbook solutions, approved lessons learned, programs of instruction, hardware, and other established, by the book structures.

Educating future strategists in this paradigm would involve implementing the following sorts of practices. Engage practitioners in investigating organizational capacities and abilities to train and perform well-defined roles, missions, and pre-defined tasks (e.g., studying military technical capabilities, applying the Joint Chiefs of Staff Manual [CJCSM], *Universal Joint Task List* [CJSCM 3500.04], and so on). Educators also involve practitioners in processing historic case studies to determine the causes of successes and failures of past strategies. Curricula are designed to have students look for logical match-ups between roles and tasks and the effectiveness of these *cogs* in the *engine* of change: doctrine, training, materiel, leadership, personnel systems, facilities, etc. The idea behind programmatic strategy is to then routinize those causes (institutionally labeled "lessons learned") by institutionalizing them into machine-like routines that will be called upon later.

STRATEGIC REASONING THROUGH PLANNING

Theory of Reasoning.

Another ends-based rational approach involves planning, or "formalized procedure to produce an articulated result, in the form of an integrated system of decisions" that are interpreted by policymakers as being important to future success.[8] Planning is associated with convergent knowledge in that it involves a reasoned approach to excluding other possible courses of action. Planning addresses more intractable situations where known technologies are inadequate to define the problem and where *convergent knowledge* (i.e., discovery of viable courses of action by synthesizing known-knowns and excluding alternatives) is prevalent. The degree of complexity of the situation is aligned with an equal complexity of solutions (see Figure 3).

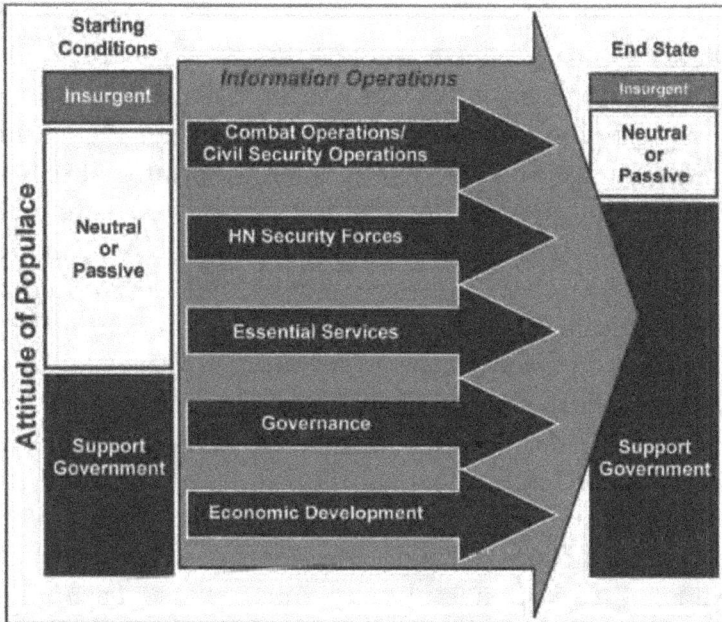

Figure 3. A Graphic Example Of Strategic Reasoning Through Planning.

Strategy is depicted as "logical lines of operations for a counterinsurgency" on page 5-3 of the U.S. Army *Field Manual (FM) 3-24/Marine Corps Warfighting Publication 3-33.5, Counterinsurgency*. Note that both analysis (conditions, lines of operation, end state, etc.) and synthesis (the government activities are blended to affect the *attitude of populace*) can be detected from this diagram.[9] Success (defined by the end state) is attributed to these logical lines retrospectively.

This view of strategy assumes more complexity than the programmatic view, yet still relies much on determinism.

Like programmatic reasoning, reasoning-through-planning involves understanding historic precedent. In addition, planning requires modifications to precedent based on the degree of intractability of the situation. Defining the problem cannot be clearly or simply linked to a known technology. The strategic planner looks for new ways to combine technologies over time. Hence the situation becomes a problem when the strategist sees it as amenable to his composition of selected solutions over time. The complexity of the problem becomes more definitive with the creation of blended-solutions. The technological emphasis shifts from matching (the emphasis in strategic programming) to "bricolating," where the latter involves the more inventive processes of kluging solutions.[10] Because there is an inherent risk involved in kluging solutions (i.e., the uncertainty as to whether the concoction will work), the planner takes precautions through the creation of contingency plans. A useful metaphor to understand reasoning-through-planning are the techniques and aesthetics of composers of orchestra music. Of all the possible combinations and permutations of musical notes, the composer converges on relatively few; they sound good, they resonate, and are aesthetically pleasing.

A recent example of strategic planning that worked is the 1990-1991 Gulf War.[11] By the end of the 1980s, the United States had nearly perfected its programmatic strategy (the orchestration of doctrine, organization, training, transportation system, and so forth) and then bricolaged its forces with an array of international forces to remove Iraqi forces occupying Kuwait. The *orchestration* (the dominant metaphor for this example) required to *conduct* the campaign was accomplished through detailed planning, in which the ends were conveniently articulated clearly by the policymakers. The military strategists could "bricolate forces" and *harmonize* them with other *instruments* of power to accomplish them (all activities were *played* on the same *sheet of music*).

Implications For PME.

Insofar as PME andragogy is concerned, educators seek to educate practitioners from the point of view of the planning strategy paradigm, relying upon convergent knowledge, where abstract concepts are transformed into probabilistic, realizable goals and objectives that can lead to technical comprehension (i.e., knowledge that is preparatory to getting to the desired tractability and clarity afforded by more programmatic strategies). Convergent knowledge in this form of reasoning tends to be an institutionalized process of ends-based rational decisionmaking (e.g., the Joint Strategic Planning System). Educators under the planning paradigm (associated with convergent knowledge) would involve practitioners with variations on the following generic process steps: (1) *Initiate* — Receive and understand the policy decision; (2) *Estimate* — Refine the problem definition, define tasks and set objectives; then search for organizational courses of action (the plan "converges") to accomplish tasks and meet objectives; (3) *Select* — choose the best course of action; and, *Implement, Evaluate, Terminate* — tasked organizations execute until missions are complete and objectives are met. [12]

Variations to this deliberation include adding substeps and parallel steps that can create very sophisticated planning repertoires. [13] In that regard, military planners in particular have recognized that refining the problem definition can be done in more complicated sets or ranges of problems (e.g., combat + instability + humanitarian crises, and so on) that can address tasks that may involve a wide variety of kluging organizations and activities (service + joint + interagency + international + nongovernmental, and so on). This paradigm overlaps with the programmatic paradigm in that regard, and separate organization routines can be designed with modularity, capable of adapting organizational building blocks toward matching the complexity of the solution with the complexity of the defined problem.[14] Instruction methods can range from practicing planning with respect to well-developed scenarios to reviewing historic case studies where burgeoning strategic planners can compare planned events with implementation, revealing aspects failure, success, and best practices of risk mitigation to counter the uncertainty involved.

STRATEGIC REASONING THROUGH PARTICIPATION

Theory of Reasoning.

Participative forms of strategic reasoning rely on the relatively unstructured process of accommodating multiple ways of contextualizing situations. Methods of strategy here represent the antithesis of programmatic or planning ways of thinking. In this paradigm, meaning is negotiated through interpretation and social interactions. Because the situation is highly ambiguous (i.e., there are multiple interpretations of what is going on and what to do about it), participative reasoning requires the understanding of multiple viewpoints and trying to shape them into a consensus or at least appreciate the intractability of reaching consensus. This may involve lengthy dialogue, diplomacy, deception, coalition-building, negotiating, use of propaganda, confrontation, and other forms of participative decisionmaking.

Both here and with the reflective paradigm (addressed next), there "is no such thing as a logical method of having new ideas" or agreeing to them.[15] In the process of negotiating shared meaning, the participative paradigm may shift to one or more of the other three types of reasoning (see Figure 4). The dominant metaphor for this mode of reasoning is the pluralistic form of politics, emphasizing these concepts: sense of community; sense of common interests and problems; the paradox of cooperation and competition; groups and organizations are "building blocks;" "information is interpretive, incomplete, and strategic;" and, where the "laws of passion" may trump the laws of physics.[16]

	Agreement on Means/Ways	Disagreement on Means/Ways
Agreement on Ends	Move to **Programmatic** mode If causality is certain, or to **Planning** if uncertain	Stay in **Participative** Mode
Disagreement on Ends	Move to **Planning** mode (Does it really matter? Just do it!)	If no agreement in sight consider that you may be "stuck" in the **Reflective** mode

Figure 4. Diagnosing Accommodation.

Multiple interpretations of what is going on and what needs to be done define the ambiguous nature of strategy under the participative paradigm (in other words, "politics"). The strategist is best served by diagnosing progress by discerning which form of reasoning seems to be more applicable to approach accommodation. This is not to say that randomness, hidden-agendas, guile, and other "Machiavellian" aspects of negotiating a strategy do not come into play; so, "ends-based rationality" may still play a deceptive premise for preset agendas and decisionmaking.[17]

A classic example of a pluralistically reasoned strategy that eventually worked was documented by British Field Marshal (then Lieutenant General) William Slim. In his book, *Defeat into Victory*, Slim wrote about how he managed to successfully win in the China-Burma-India Theater during World War II, after the Japanese soundly defeated his coalitional Burma Corps. He reflected on the ambiguity of the 1942-Allied retreat into India:

> To me, thinking it all over, the most distressing aspect of the whole disastrous campaign had been the contrast between our generalship and the enemy's. The Japanese leadership was confident, bold to the point of foolhardiness, and so aggressive that never for one day did they lose the initiative. True, they had a perfect instrument [an army highly trained and equipped for joint jungle warfare] for the type of operation they intended, but their use of it was unhesitating and accurate. Their object, clear and definite, was the destruction of our forces: ours a rather nebulous idea of retaining territory. This led to the initial dispersion of our forces over wide areas, an error which we continued to commit, and worse it led to a defensive attitude of mind...it was painfully obvious that the lack of a definite, realistic directive

from above made it impossible for our immediate commanders to define our object with the clarity essential...we had been weakened by this lack of a clear object.[18]

Later in the book, Slim made this paradoxical statement about a strategy-of-not-discussing-strategy (exemplifying the political logic of participative mode of reasoning):

> ...Admiral Mountbatten's staff...realized clearly that Stilwell was very much the senior American general . . . suddenly . . . Stilwell astonished everyone by saying, "I am prepared to come under General Slim's operational control. . . . [This] created an even more illogical situation. Luckily he and I were determined on the same things—to get more Chinese divisions for the Ledo force, to push hard for Myitkyina, and to use Wingate's Chindits to aid that push. . . . Tactically we were in agreement, wisely, we avoided strategic discussion.[19]

Implications for PME.

The participative mode is most linked to accommodative knowledge that requires flexibility of thought (e.g., temporarily suspending disbelief in other ways to frame or describe the situation at hand) while accepting more unstructured and intangible ways of negotiation.[20] Unstructured strategy making may be defined as "decision processes that have not been encountered in quite the same form and which no predetermined and explicit set of ordered responses exists in the organization" or among the institutions represented.[21] This sort of unstructured strategy making then is a groping or muddling through, messy, and recursive process that requires a certain patience for participative forms of reasoning and a tolerance for building accommodation. Instead of a comprehensive approach to strategy, the resulting strategy becomes a series of successive limited comparisons. For very novel situations (such as those uniquely encountered in countering insurgencies) the strategic reasoning method is "muddling through" — an ill-structured series of incremental recontextualizations.[22] In other words we act, assess, react, and so forth, comparing the situation now to what it previously was to look for "improvements." Whether these comparisons reflect improvement is a socially negotiated consensus process.

Educating strategists under this paradigm involves appreciating conflicting values and the associated variety of individual, group, and societal emotions, cultural proclivities, ethical positions, aesthetic feelings, and religious beliefs, as well as the interpretations that stem from them. Making situations seem more tractable may include treating social groups and nations as if they were unitary actors — attributing individual human-like motivations and decisionmaking capacities to groups, organizations and nations. Methods of student inquiry should include hermeneutics — a kind of historic accounting that uses the humanities versus the natural sciences' or positivist approach. This view requires educators to help practitioners critically examine human communications in the search of deeper understanding. For example, hermeneutics would demand not just regurgitating what another person or group expresses, but also an attempt to interpret the thinking behind it, perhaps framed from an entirely different worldview.[23]

Hermeneutics is related to the idea of interactive learning, which explores how human actors improve and create shared meaning through social relations. Here, strategy making is assumed to be a relatively free-flowing, socially interactive process. The exploration of that unstructured process requires methods that are "flexible, imaginative, creative, and free to take new directions."[24] Philosophical sociologist Herbert Blumer describes methods to study social interactions as antithetical to natural scientific methods.[25] This qualitative method of critical inquiry involves the practitioner in being alert to the need of testing and revising images, beliefs, and conceptions that would otherwise distort understanding. Like an ethnographer, being able to *richly describe* what is interesting (even if confusing at the time) is important to the practice "even if its relevance is not immediately clear."[26]

Exercising these interpretive, critical, and imaginative skills involves the practitioner in exploring deliberate attempts to change other actor's thinking and behavior through political guile or sanctions: inducements (incentives and penalties); rules (mandates); facts (informing and persuading); rights (and duties); and powers (authority) where " . . . 'new policies' are really somebody's next move."[27] The practice of political decisionmaking is so nuanced and unstructured as to better be developed in actual situations than in the classroom.[28] Based on research in adult experiential learning, educating the practitioner involves field work, perhaps through apprenticeships and practicum with the other agencies, international exchange programs, or the like.[29] There is no substitute for on-the-scene experience; albeit, such experience may not be transferable to other situations. The real value of the "real world" immersion is that the practitioner may become more comfortable in the unstructured reasoning processes within the uniqueness of each condition at hand.

STRATEGIC REASONING THROUGH REFLECTION

Theory of Reasoning.

Strategic reasoning under conditions of high ambiguity and intractability (also characterized as "messes" or "wicked problems") is characterized by the relatively unstructured process of practitioner sensemaking while being mindful of his own limitations.[30] In strategic reflection, *divergent knowledge* is required as practitioners come to a realization that they face large-scale, complex, or chaotic situations where institutionalized knowledge is insufficient or nonexistent. Such reasoning becomes the unstructured process of sensemaking, through abductive reasoning, in highly complex and subjective situations, while reflecting critically on an opaque awareness that there are many "unknown unknowns."[31]

Indications that reflexive forms of reasoning should prevail occur when there is the realization that situations are more than complicated and complex than "problems"; are highly ambiguous; contain considerable uncertainty—even as to what the conditions are, let alone what the appropriate actions might be; are tightly interconnected—economically, socially, politically, and technologically—and, appear paradoxical.[32] Here there is no chance of routine application of professional knowledge because practical knowledge will have to be invented as we go (i.e., divergent forms of knowledge are required). New rules to govern inquiry have to be created in the face of anarchical situations and then those too will have to be also questioned as to whether the new way of reasoning seems to be working. For the reflective practitioner, these novel situations reveal "indeterminate zones of practice" (also known as artistry).[33] The dominant metaphor for reflective reasoning is the "improvisational art;" hence, whereas planning is associated with orchestration, reflection may be better associated with improvisational jazz.

It is difficult to find historic examples of strategizing under these conditions because, in retrospect, we now know how things turned out (so the "wickedness" of the milieu experienced then is difficult to articulate now). Perhaps the ongoing example of how the United States and the North Atlantic Treaty Organization (NATO) have adapted their strategy since 2001 in Afghanistan presents a worthy case because the wickedness seems even more prevalent presently than it was in the early years of the U.S. invasion. Conceptualizations of ends, ways, and means seem in continual states of flux and transformation. While it seems as though the problem has been the lack of a thoughtful and workable ends-based strategy, reflection involves questioning ways of reasoning about the situation that do not seem to work and developing ways to reframe.

Implications for PME.

Under these sorts of highly intractable and ambiguous conditions, the educator's role is to convince practitioners that they would benefit from Donald Schön's concept of *reflective practice*—treating the seminar or small group learning forum as a "design studio," as a class in "musical conservatory," or as "counseling-in-action." The critical function of reflection-in-action, according to Schön, is "questioning the assumptional structure of knowing-in-action."[34] That assumptional structure is called "framing." Indeed, Schön substantiates that educating practitioners to be reflective requires that the student, ". . . think critically about the thinking that got us into this fix or this opportunity; and we may, in the process, restructure strategies of action, understandings of phenomena, or ways of framing. . . ."[35] This restructuring of sensemaking reflects the creative essence of divergent knowledge and has improvisational qualities not associated with ends-based rationality.[36] Indeed, situations that are not conducive to strategy may require an ongoing process of inventiveness.[37] Such messes require constant *resolutions* (i.e., more analogous to managing symptoms of a chronic illness, like AIDS, than attempts to cure the patient).[38] Practitioners must become good at reflective reasoning while actions are underway (i.e., they must reflect-in-action).

The artful framing and reframing of situations is inherently important to the creative and recursive design of strategy-as-we-go. Here, students are engaged in the reflective practice of strategic design, uniquely, and continuously crafted "within the context and tailored to fit some conception of the situation."[39] Strategists realize that strategic activities are not just a result of a political process, but interactively cause politics — that is, the concept of *mutual causality* applies.[40] This perspective complexifies the Clausewitzian principle that asserts war is policy by other means and recognizes that activities in war may cause policy changes (as we have witnessed in Afghanistan and, more recently, reframed as the Afghanistan-Pakistan area of concern).[41]

Furthermore, educators must tap into practitioner appreciations of complexity science, chaos theory, and social construction theory.[42] Complexity science and chaos theory facilitate practitioners' language for appreciating many interactive variables within social systems. To appreciate the complexity and the mutual causality inherent to Taliban attacks in Afghanistan, one might map the complicated interrelationships as shown in Figure 5.

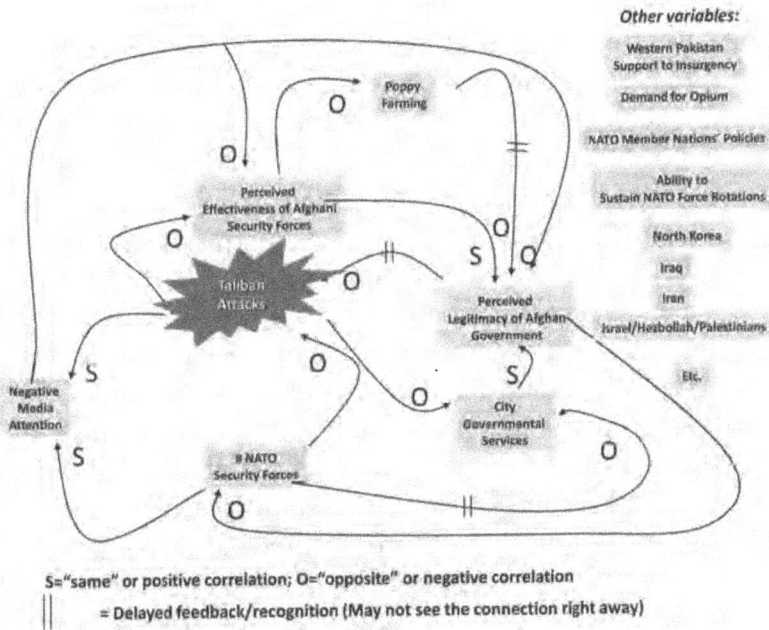

S="same" or positive correlation; O="opposite" or negative correlation

|| = Delayed feedback/recognition (May not see the connection right away)

Figure 5. Causal Loop Diagram of the Afghan-Pakistan Situation.

Causal loop diagrams are used to portray complex relationships among interacting variables to the point one realizes that predictability of the "system" or its "parts" is not possible (why Herbert A. Simon called this sort of modeling the "science of the artificial" and "simulation of poorly understood systems").[43] As the Taliban inflicted violence goes up, the perceived effectiveness of Afghan security forces goes down. As numbers of NATO security forces increases, Taliban sources of violence may go down, but so may the positive perception of Afghan security forces . . . and so on. Looking at all of these interacting variables and noting that there may be delayed feedback loops, one can only appreciate (not fully understand) the idea of "nonlinear" or "dynamic" feedback.[44] The idea of side effects or unintended consequences is more appreciable, yet still unpredictable.

Chaos theory would suggest that manipulating one or more variables may unpredictably amplify significant changes within the system (i.e., the butterfly effect). In view of the complexity and chaotic nature of social systems, social construction theory provides explanations as to how social groups interact to invent sources of meaning where objective explanations are not plausible.[45] In that regard, designing strategy postulates that "there is no answer that should be considered 'true' for all times and places, but that through discourse a more limited and contingent type of truth may emerge" (hence the proverb, "the truth changes").[46] Therefore, dialogue, collaborative inquiry, and action research become the tools that PME educators must stress with practitioners. These tools enable them to collectively make better sense of otherwise intractable and ambiguous situations. This sensemaking *is* strategic reasoning (and sensemaking is a contextual precursor to strategy making).

SPECULATIONS AND CONCLUSION

This chapter has asserted that four kinds of reasoning (programmatic, planning, participative, and reflective) have to be considered simultaneously—a talent at being quad-conceptual (that is, the capacity and ability to think in patterns). With the proposed multifaceted framework for appreciating the difficulties of educating future strategists in the midst of social messes and wicked problems (such as famine, war, poverty, failed economies, and so on), it is hoped that educators can at least help practitioners of strategy to appreciate the simultaneity of these four perspectives on strategic reasoning (and perhaps contemplate additional paradigms as they further inquire). Each type provides a unique and valuable perspective on the nature of reality and the type of knowledge constructions that go with those reasoning skills. Rather than display these as irreconcilable paradigms, the framework presented in this chapter seeks to transcend the otherwise incommensurability of opposites. For example, here are four patterned archetypes that reflect the author's interpretation of the simultaneity of degrees of tractability and ambiguity of situations that were used as examples in this chapter (see Figure 6). While there may be an overall or predominant perspective on reasoning (indicated by the tail of the kite in the patterns) there are some aspects of the other types concurrently in use.

participative	reflective
programmatic	planning

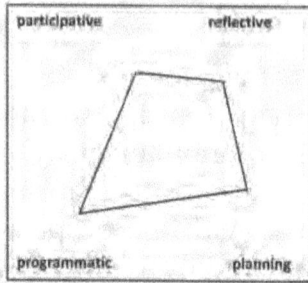

Force Projection Strategy

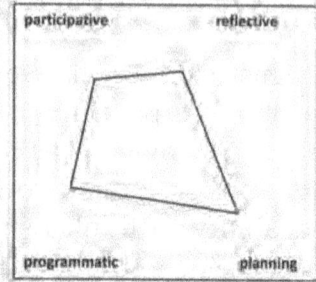

participative	reflective
programmatic	planning

1990-91 Gulf War

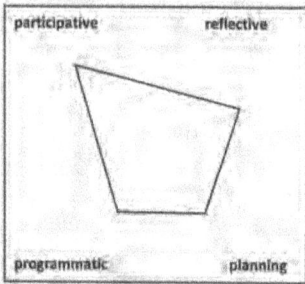

participative	reflective
programmatic	planning

China-Burma-India Theater, WW II

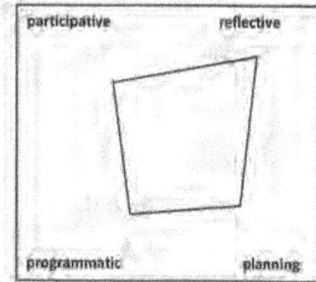

participative	reflective
programmatic	planning

Afghanistan-Pakistan Situation

Figure 6. Quad-Conceptualizations of Strategic Reasoning.

Using example situations explained earlier in the chapter, the author suggests that patterns reveal that all four forms of reasoning can be at work simultaneously. These patterns of strategic thinking reflect a snapshot-in-time, demonstrating some aspect of each paradigm is present all of the time. Patterns will shift unpredictably and (like snowflakes) no two patterns will be the same.[47] Some situations are interpreted as less ambiguous than others because we can reason them primarily through programmatic and planning processes (lower quadrants). Other situations are more associated with high ambiguity (upper quadrants). Dominance in the programmatic and planning quadrants reflect more ends-based rational forms of reasoning associated with less ambiguous circumstances, while dominance in the participative and reflective quadrants indicate situations that are not as amenable to ends-based rationality and require ongoing negotiation, collaboration, critical reflection, and a highly improvisational mindset.

Based on experience as a faculty member at the USAWC and the U.S. Army Command and General Staff College, the author speculates that staff and war college curricula today are overly geared to ends-based rationality. The ideology of ends-based rationality must be subordinated to a more comprehensive model of reasoning—inclusive of the underemphasized ideas of participative and reflective thinking. Practitioners may have to design accommodative and divergent forms of knowledge to frame, reframe, make sense, negotiate realities about novel, highly complex, and ambiguous situations, and reflect-in-action.

Finally, how would one design curricula for strategic reasoning using the quad-conceptual model? How should we prepare faculty to facilitate the curriculum? Traditional curricula overemphasize the development of reasoning for programmatic and planning strategies. One reason may be that these forms of reasoning can be readily engineered into curricula based in known-knowns (such as military capabilities) and by exercising rational decision and planning processes. Participative and reflective forms of reasoning are best developed through real world experience because it is implausible or impossible to replicate novel, volatile, uncertain, complex, and ambiguous situations in the classroom where the stakes are not high. Practicum and apprenticeships offer the best approach to andragogy, where the faculty is developed to serve as coach and mentor, helping the apprentice shape these unstructured reasoning skills in tandem with the leadership in those agencies that provide opportunities for practicum and apprenticeship. Formalizing practicum and apprenticeships in PME and in the defense, interagency, intergovernmental, and multinational communities may take major reform efforts.

ENDNOTES - CHAPTER 11

1. Nicholas Rescher, *Luck: The Brilliant Randomness of Everyday Life*, Pittsburgh, PA: University of Pittsburgh, 1995. Rescher's central thesis is that projecting future events is impossible because of the interactions of chance, chaos, and choice.

2. This argument about actuality with respect to historicity is well documented by Alfred Schutz in his Collected Papers, Vol. II, *Studies in Social Theory*, The Czech Republic: The Hague: Martiners Nijhoff, 1964, pp. 56-57.

3. The idea of "quad-conceptual" (*thinking-in-fours*) comes from the work by Sabrina Brahms, *Systemic Narrative Inquiry (SNI) Method: Theory Presentation and Application within the Israeli-Palestinian Conflict*, Proceedings of the 47th Annual Meeting of the International Society for the Systems Sciences at Hersonissos, Crete, July 6-11, 2003, available from *systemicbusiness.org/digests/isss2003/2003_ISSS_47th_079_Brahms.pdf*, p. 3.

4. For this chapter, technology is defined as: ". . . all the [assimilative] knowledge, information, material resources, techniques, and procedures that a work unit uses to convert system inputs into outputs — that is to conduct work." Rupert F. Chisholm, "Introducing Advanced Information Technology into Public Organizations," *Public Productivity Review*, Vol. 11, No. 4, 1988, pp. 39-56.

5. Deborah Stone, *Policy Paradox: The Art of Political Decisionmaking*, 2d Ed., New York: W. W. Norton & Company, 1997. Stone euphemistically calls this reasoned approach the

"rationality project," p. 6. In PME institutions, the ends-based rationality is called "instrumental rationality," where *instruments of power* (diplomatic, informational, military, economic, etc.) are arrayed toward ends.

6. Donald A. Schön, *Educating the Reflective Practitioner*, San Francisco: Jossey-Bass, 1987, p. 33.

7. See Christopher R. Paparone and George E. Reed, "The Reflective Military Practitioner: How Military Professionals Think in Action," *Military Review*, Vol. 88, No. 2, 2008, pp. 66-76. The authors discuss the relationship between these Kolbian knowledge types (assimilative, convergent, accommodative, and divergent) and how the professional knowledge process works. Kolb's multiple epistemological views are remarkably close to the arguments made in the present paper, so it should not be surprising that these types match quite well to the four quadrants presented. Given all the knowledge types, the authors argue that *reflective practice* is a superior approach to the more traditional search for *best practice*. In the context of this chapter, the quest for best practice is stuck in the programmatic and planning paradigms.

8. Henry Mintzberg, *The Rise and Fall of Strategic Planning: Reconceiving Roles for Planning, Plans, Planners*, New York: Free Press, 1994, p. 12.

9. See also the recently released *Field Manual (FM) 3-07, Stability Operations*, October 6, 2008, which appears to be an attempt at "whole government" doctrine. The FM defines whole government as "an approach that integrates the collaborative efforts of the departments and agencies of the United States Government to achieve unity of effort toward a shared goal," p. 1-4. Nevertheless, to this writer, "whole government" is a misleading term in that the important legislative branch is not included. In a partisan system of government, will there ever be a "whole government" approach? The author doubts such an approach is possible or even desirable in the dynamic politics of a republican democracy.

10. The term "bricolage" was coined with this meaning by Claude Lévi-Strauss in his book, *La Pensee Sauvage, The Savage*

Mind. Bricolage emphasizes resilience by forming new ways to accomplish things through the creative use of existing knowledge. Paradoxically, the improvised use of assimilated knowledge (called convergent knowledge) can be quite creative and result in a new knowledge creation in itself. See Karl E. Weick, "Improvisation as a Mindset for Organizational Analysis," *Organization Science,* Vol. 9, No. 5, 1998, pp. 543-555.

11. The types presented are ideal, so it is difficult to find pure examples. In reality, all forms of strategic reasoning are required. It is a matter of degree. Gulf War history demonstrates aspects of the other three types of reasoning as well; yet, *planning* appeared to be the preeminent form of reasoning for action.

12. Peter DeLeon, "The Stages Approach to the Policy Process," in Paul A. Sabatier, ed., *Theories of the Policy Process: Theoretical Lenses on Public Policy,* Boulder, CO: Westview, 1999, pp. 19-22.

13. Note, when hyphenated, the word "de-liberation" means literally to take away freedom of action; and, in this case, to focus resources on a single course of action, that trumps all others.

14. See Melissa Schilling and Christopher R. Paparone, "Modularity: An Application of General Systems Theory to Military Force Development," *Acquisition Review Journal,* Vol. 12, 2005, pp. 279-293.

15. This is an adaptation of Karl Popper's argument in his book, *The Logic of Scientific Discovery,* London, UK: Routledge, 2002 (first published in 1935), p. 8. The present author extends this argument to achieving negotiated agreements, i.e., a very creative process that relies on experiential learning.

16. Stone, p. 32.

17. This chart was adapted by the author from the ideas of James D. Thompson and Arthur Tuden, "Strategies, Structures, and Processes of Organizational Decision," in J. Thompson, ed., *Comparative Studies in Administration,* New York: Garland, 1987, pp. 195-216. (Original work published in 1959.)

18. Field Marshal The Viscount Slim, *Defeat Into Victory*, New York: David McKay Company, 1961, pp. 95-96.

19. Slim, pp. 179-180.

20. The Greek term for this suspension of disbelief is *epochè*.

21. Henry Mintzberg, Duru Raisinghani, and Andre Theoret, "The Structure of 'Unstructured' Decision Processes," *Administrative Science Quarterly*, Vol. 21, 1976, p. 246.

22. Charles E. Lindblom, "The Science of Muddling Through," *Public Administration Review*, Vol. 19, No. 2, Spring, 1959, pp. 79-88.

23. Michael D. Pearlman, *Warmaking and American Democracy: The Struggle over Military Strategy 1700 to the Present*, Lawrence: University Press of Kansas, 1999.

24. Herbert Blumer, *Symbolic Interactionism: Perspective and Method*, Berkeley: University of California, 1969, p. 44.

25. *Ibid*. Traditional scientific methods are considered by Blumer to operate "unwittingly with false premises, erroneous problems, distorted data, spurious relations, inaccurate concepts, and unverified interpretations," p. 29.

26. *Ibid*, p. 42. Also see Karl E. Weick, "The Generative Properties of Richness," *Academy of Management Journal*, 2007, Vol. 50, No. 1, pp. 14-19. Weick states that the argument for better rich description "is an argument for detail, for thoroughness, for prototypical narratives, and an argument against formulations that strip out most of what matters. It is an argument that the power of richness lies in the fact that it feeds on itself in ways that enlarge our understanding of the human condition," p. 19.

27. Stone, 1997, pp. 13, 259. Furthermore, Stone defines policy as the "rational attempt to attain objectives . . . by interpreting, through political reasoning, the criteria for choice," pp. 37-38. She argues those criteria (justifications for policy or values) fall into four major value categories: equity, liberty, efficiency, and security, p. 37.

28. Perhaps this explains why the State Department does not invest in professional schools for its foreign service officers as does the DoD for its officers' "professional military education."

29. For example, see David A. Kolb, *Experiential Learning: Experience as the Source of Learning and Development*, Englewood Cliffs, NJ: Prentice-Hall, 1984.

30. For "messes," see Russell L. Ackoff, *Ackoff's Best: His Classic Writings on Management*, New York: Wiley, 1999. Ackoff defines messes as "systems of problems, [where] they lose their essential properties when they are taken apart," p. 117. For wicked problems, see Horst W. Rittel, H. and Melvin M. Webber, "Dilemmas in a General Theory of Planning," *Policy Sciences*, Vol. 4, 1973, pp. 155–169. Wicked problems are at the other end of the spectrum from tame problems, hence, better characterize problems of government policy. According to Rittel and Webber, wicked problems are "incorrigible" that "defy efforts to delineate their boundaries and to identify their causes, and thus to expose their problematic nature," p. 167.

31. Atocha Aliseda, *Abductive Reasoning: Logical Investigations into Discovery and explanation*, Dordrecht, The Netherlands: Springer, 2006. According to Aliseda, abductive reasoning involves the discovery of tentative inferences and search strategies for possible explanations. The viability of these explanations is affected by luck, persistence, and superior heuristics (the latter stemming from past experience, imagination, metaphors, and other forms of analogical reasoning). Surprise is the trigger of abductive reasoning; hence, goes hand-in-hand with doubt and the impetus to change a belief system. The motivation for discovery is to sooth (as in soothsaying/fortune-telling) and to create suggestions that must be bested before converting into belief.

32. Adapted from Robert E. Horn, "Knowledge Mapping for Complex Social Messes," A presentation to the "Foundations in the Knowledge Economy," the David and Lucile Packard Foundation, July 16, 2001, May 22, 2006, available from *www. stanford.edu/~rhorn/images/SpchPackard/spchKnwldgPACKARD.pdf*. Horn was interpreting the work of Russell Ackoff and came up

with the idea of creating graphic views of messes through cross-boundary causality maps.

33. Schön, p. 39.

34. *Ibid.*, p. 28.

35. *Ibid.*

36. Weick, pp. 543-555. Qualities of improvisation include dealing with the unforeseen and unexpected that may include "flexible treatment of preplanned material." Other qualities include playing extemporaneously, making a difference with ongoing action, spontaneity, novel activities, creation of something while it is being performed, is linked to intuitive processes, requires a "disciplined practicer," p. 544. "Wade in and see what happens," p. 548. "Composing on spur of the moment," p. 548.

37. Ackoff, p. 115.

38. *Ibid.*

39. Anne Larason Schneider and Helen Ingram, *Policy Design for Democracy*, Lawrence: University of Kansas, 1997, p. 69.

40. *Ibid.*, p. 74.

41. Alan Beyerchen, "Clausewitz, Nonlinearity, and the Unpredictability of War," *International Security*, Vol. 17, No. 3, Winter 1992-93, pp. 59-90.

42. Robert L. Flood and Norma R.A. Romm, eds., *Critical Systems Thinking: Current Research and Practice*, New York: Plenum Press, 1996. Also see Peter L. Berger and Thomas Luckmann, *The Social Construction of Reality: A Treatise in the Sociology of Knowledge*, New York: Anchor, 1967.

43. Herbert A. Simon, *The Sciences of the Artificial*, 3d Ed., Cambridge, MA: MIT, 2001, p. 5.

44. Peter Senge maintains that causal loop diagrams allow us to see "the 'structures' that underlie complex situations, and

for discerning high from low leverage change. That is by seeing wholes we learn how to foster [system] health." Peter Senge, *The Fifth Discipline: The Art and Practice of the Learning Organization*, New York: Doubleday Currency, 1990, p. 69. In complex systems, *"doing the obvious thing does not produce the obvious, desired outcome,"* emphasis in original, p. 71. Also see Geoff Coyle, *Qualitative Modeling in System Dynamics or What are the wise limits of quantification?* A Keynote Address to the Conference of the System Dynamics Society, Wellington, New Zealand, May 1999, available from *www.systemdynamics.org/conferences/1999/ PAPERS/KEYNOTE1.PDF*; and Edward G. Anderson Jr., "A Proof-of-Concept Model for Evaluating Insurgency Management Policies Using the System Dynamics Methodology," *Strategic Insights*, Volume VI, Issue 5, August 2007, available from *www. nps.edu/Academics/centers/ccc/publications/OnlineJournal/2007/Aug/ andersonAug07.pdf.*

45. One of the ways social groups "invent" these meanings is to *borrow* meaning from other knowledge disciplines: natural sciences, social sciences, and the humanities. For a treatise on this subject, see Christopher R. Paparone, "Metaphors We are Led By," *Military Review*, November-December 2008, pp. 55-64.

46. Schneider and Ingram, p. 62.

47. Simon, 2001, p. 33.

ABOUT THE CONTRIBUTORS

THOMAZ GUEDES DA COSTA is a Professor of National Security Affairs, Center for Hemispheric Defense Studies at the National Defense University. Previously, he served at: Brazil's National Council for Scientific and Technological Development; as an analyst at EMBRAER; as a researcher/advisor in the Center for Strategic Studies, Secretariat of Strategic Affairs, at the Office of the Brazilian Presidency; and taught at the Department of International Relations of the University of Brasilia. Dr. Costa has written extensively on Brazil and strategic issues. His publications include "Engaging a Rising Brazil," "Brazil's SIVAM: As It Monitors the Amazon Will It Fulfill Its Human Security Promise?" and *Brazil in the New Decade: Searching for a Future*. Dr. Costa holds a Ph.D. in Political Science from Columbia University.

ROBERT H. "ROBIN" DORFF joined the Strategic Studies Institute in June 2007 as Research Professor of National Security Affairs, and currently holds the General Douglas MacArthur Chair of Research at the U.S. Army War College (USAWC). He previously served on the USAWC faculty as a Visiting Professor (1994-96) and as Professor of National Security Policy and Strategy in the Department of National Security and Strategy (1997-2004), where he also held the General Maxwell D. Taylor Chair (1999-2002) and served as Department Chairman (2001-04). Dr. Dorff has been a Senior Advisor with Creative Associates International, Inc., in Washington, DC, and served as Executive Director of the Institute of Political Leadership in Raleigh, NC (2004-06). Dr. Dorff remains extensively involved in strategic leadership development, focusing on national security strategy and policy, and strategy formulation. His research interests include these topics as well as failing and fragile states, interagency processes and policy formulation, stabilization and reconstruction operations, and U.S. grand strategy. He lectures frequently on these topics and has spoken all over the United States and in Canada, Europe, Africa, and Asia, and at institutions such as the Africa Center for Strategic Studies, the Near East-South Asia Center for Strategic Studies, the George C. Marshall Center, the Marine Corps University, the Joint Special Operations School, the National Defense University of Taiwan, and the Institute of Defence and Strategic Studies in Singapore. He is the author or co-author of three books and numerous journal articles. Dr. Dorff holds a B.A. in Political Science from Colorado College and an M.A. and Ph.D. in Political Science from the University of North Carolina-Chapel Hill.

STEPHEN O. FOUGHT is Professor Emeritus and former Dean of Academic Affairs at the Air War College. He also served for 18 years at the Naval War College as Professor, Director of Electives, and Forrest Sherman Chair of Public Diplomacy. Dr. Fought is the co-founder of the Teaching Strategy Group and has lectured on U.S. national security policy throughout the United States, Europe, and South America over the past 25 years. In addition to the chapter in this volume with Dr. Marcella (based on an article in January 2009 issue of *Joint Force Quarterly*), his other recent publications include "The Tale of the C/JFACC: A Long and Winding Road," USAF *Air and Space Power Journal* (Winter 2004), and the *RAF Air Power Journal* (Vol 7, No. 4); and "The War College Experience," feature article in *Academic Exchange Quarterly* (Summer 2004).

Dr. Fought holds a B.S. from the Georgia Institute of Technology, an M.S. from the University of Southern California, and a Ph.D. from Brown University.

VOLKER C. FRANKE is Associate Professor of Conflict Management at Kennesaw State University. From 2006 to 2008, he served as Director of Research at the Bonn International Center for Conversion, one of Germany's premier peace and conflict research and capacity building institutes. Currently, Dr. Franke serves on the Scientific Advisory Board of the German Foundation of Peace Research. He was also Director and Managing Editor of the National Security Studies Case Studies Program at Syracuse University's Maxwell School of Citizenship and Public Affairs. He is the author of *Preparing for Peace: Military Identity, Value-Orientations, and Professional Military Education* and numerous journal articles and *case* studies on peace and security issues, civil-military cooperation, social identity, and military socialization. He is also the editor of *Terrorism and Peacekeeping: New Security Challenges* and *Security in a Changing World: Case Studies in U.S. National Security Management*. Dr. Franke holds a Ph.D. in political science from the Maxwell School.

ROBERT C. GRAY is The Honorable and Mrs. John C. Kunkel Professor of Government at Franklin & Marshall College. He served as a Council on Foreign Relations Fellow in the Office of the Under Secretary of Defense for Policy and as a consultant to the Congressional Research Service. Dr. Gray was a research associate at the International Institute for Strategic Studies in London and was the North American Editor of *Defense and Security Analysis*. He has published extensively on weapons systems and arms control. Dr. Gray's current research focuses on strategy and the American academic community. He has published articles in: *Bulletin of the Atomic Scientists, Arms Control Today, Defense Analysis, Political Science Quarterly, Survival,* and *World Politics.* He has also published in the Headline Series of the Foreign Policy Association. His most recent article, "The Implications of Missile Defense for Northeast Asia," appeared in *Orbis.* Dr. Gray holds a Ph.D. in Government from the University of Texas at Austin.

ROSS HARRISON is Visiting Professor at Georgetown University, where he is chair of the international commerce and business concentration in the Master of Science in Foreign Service program. The focus of his teaching and writing is in the area of strategy, and his graduate courses encompass both national security and business strategies. Related to his work on strategy, he also has expertise in Middle East politics. Prior to his joining the faculty of Georgetown University he was Executive Vice President of Tapco International, a global building products company, and President and CEO of Builders Edge, one of the largest of the company's operating divisions. In both of these capacities, he was responsible for formulating and implementing the firm's global strategies. Currently he is researching a book on strategy.

ROBERT KENNEDY is Professor in the Sam Nunn School of International Affairs, Georgia Institute of Technology, Atlanta, Georgia, and previously served as director of the joint German-American George C. Marshall European Center for Security Studies in Germany. In nearly 35 years of government service, Dr. Kennedy has also served as Civilian Deputy Commandant, NATO Defense College, Dwight D. Eisenhower Professor of National Security Studies at the U.S. Army War College, researcher at the Strategic Studies Institute, Foreign Affairs Officer, U.S. Arms Control and Disarmament Agency, and a command pilot on active duty with the U.S. Air Force and later with the reserve forces. His most recent book is *Of Knowledge and Power: the Complexities of National Intelligence.* Forthcoming is *The Road to War: Congress's Historic Abdication of Responsibility.* Dr. Kennedy holds a Ph.D. in political science from Georgetown University

BRADFORD A. LEE has been a Professor of Strategy at the U.S. Naval War College since 1987. He was previously Assistant and then Associate Professor of History at Harvard. He has published extensively on foreign policy, civil-military relations, diplomatic history, economic theory, and economic policy. Among his publications are "Winning the War but Losing the Peace," and "Strategy, Arms and the Collapse of France, 1930-1940," in addition to *Britain and the Sino-Japanese War, 1937-1939.* He is currently preparing a major study, entitled "On Winning Wars" which will provide a new framework for understanding how military operations (along with nonmilitary instruments) translate into political results in war. Dr. Lee received his B.A. from Yale and his Ph.D. at Cambridge University and spent 3 years as a postgraduate Junior Fellow in the prestigious Society of Fellows at Harvard.

GABRIEL MARCELLA is retired Professor of Third World Studies and Director of the Americas Studies in the Department of National Security and Strategy at the U.S. Army War College. During his government career, he also served as International Affairs Advisor at United States Southern Command. Dr. Marcella has written extensively on Latin American security issues and U.S. policy. His current research focuses on the

Colombian crisis and U.S. strategy, the rule of law and governance, and national security decisionmaking. Recent publications include "American Grand Strategy for Latin America in the Age of Resentment," "Democratic Governance and the Rule of Law: Lessons from Colombia," and *Affairs of State: The Interagency and National Security*. Currently, he is adjunct professor at the U.S. Army War College. Dr. Marcella holds a B.S. from St. Joseph's University, an M.A. in History from Syracuse University, a diploma from the Inter-American Defense College, and a Ph.D. in History from the University of Notre Dame.

CHRISTOPHER R. PAPARONE is associate professor at the U.S. Army Command and General Staff College, Fort Lee Campus, Virginia. A retired Army colonel, he has held numerous command and staff positions, and his operational campaigns include foreign internal defense support to the Salvadoran armed forces, Operations GOLDEN PHEASANT, JUST CAUSE, DESERT SHIELD/STORM, and JOINT ENDEAVOR. Some of his recent publications include "Metaphors We are Led By," "The Reflective Military Practitioner: How Military Professionals Think in Action" (with George Reed), "Where Professionalism Meets Complexity Science" (with Ruth A. Anderson and Reuben R. McDaniel, Jr.), and "PPBE: A Red or a Blue Pill? Can Defense Sensemakers Really be Rational in a Hyperturbulent World?" Dr. Paparone has a Ph.D. in Public Administration from Pennsylvania State University.

CYNTHIA WATSON is Chairwoman and Professor of the Department of Security Studies at the National War College, where she has taught and held various positions since 1992. Dr. Watson pursues research on

China and Taiwan in Latin America, teaches courses on China and Latin America, and follows discussions of teaching strategy. Dr. Watson is on the Governing Board of *Third World Quarterly* and the Inter-University Seminar on Armed Forces & Society, and is also a member of the International Institute for Strategic Studies. Her publications include *Nation Building, Political Role of the Military, and National Security Groups*. She is also certified for bilingual Spanish and English at Recording for the Blind and Dyslexic. Dr. Watson earned a master's degree from the London School of Economics and a Ph.D. from the University of Notre Dame.

HARRY "RICH" YARGER is the Ministry Reform Advisor in the Security, Reconstruction and Transition Division of the Peacekeeping & Stability Operations Institute, Carlisle, Pennsylvania. He was also Professor of National Security Policy in the Department of National Security and Strategy at the U.S. Army War College where he held the Elihu Root Chair of Military Studies. His research focuses on strategic theory, national security policy and strategy, terrorism, irregular warfare, effective governance, and the education and development of strategic level leaders. In addition, he chaired the U.S. Army War College's Department of Distance Education. His latest work is *Strategy and the National Security Professional: Strategic Thinking and Strategy Formulation in the 21st Century*. A retired Army colonel, he is a Vietnam veteran and served in both Germany and Korea. Dr. Yarger is a graduate of the U.S. Army War College and holds a Ph.D. in history from Temple University.

www.ingramcontent.com/pod-product-compliance
Lightning Source LLC
Chambersburg PA
CBHW050451270326
41927CB00009B/1698